China's Creative Industries

To my family

China's Creative Industries

Copyright, Social Network Markets and the
Business of Culture in a Digital Age

Lucy Montgomery

Research Fellow, Australian Research Council funded Center of
Excellence for Creative Industries and Innovation, Queensland
University of Technology, Australia

Edward Elgar
Cheltenham, UK • Northampton, MA, USA

Published by
Edward Elgar Publishing Limited
The Lypiatts
15 Lansdown Road
Cheltenham
Glos GL50 2JA
UK

Edward Elgar Publishing, Inc.
William Pratt House
9 Dewey Court
Northampton
Massachusetts 01060
USA

A catalogue record for this book is available from the British Library

Library of Congress Control Number: 2010925995

ISBN 978 1 84844 864 3

Typeset by Cambrian Typesetters, Camberley, Surrey
Printed and bound by MPG Books Group, UK

Contents

Foreword: Whose creative industries?

John Hartley[1]

The context in which this book was produced is important, because there's a great deal more going on here than a simple observation of recent developments in an emergent economy. Lucy Montgomery has synthesized interdisciplinary approaches to the creative economy, media studies, copyright law and 'area' studies.

In an era of hyper-specialization it is unusual to find such a range of concerns, but at the same time it is impossible to do justice to specialist domains without knowing how they fit together, and how different national economic and cultural systems interconnect in a globalized economy and technologically networked culture.

CONTEXT

At QUT we have been among the pioneers of international research in the creative industries. In fact we launched the world's first Creative Industries Faculty in 2001, of which I was the foundation dean. But to achieve a preeminent research position we had to put together a team that drew on expertise in five separate disciplinary domains – media and cultural studies, law, business and economics, IT and computer science, and education.

In 2005 we won funding from the Australian Research Council for a Centre of Excellence based on this interdisciplinary mix, and went to work on a programme of research designed to:

1. account for the *creative economy* (what's special about it, economically, if anything?);
2. investigate various *creative sectors*, ranging from popular TV and film to fashion (what impediments are there on the pathway from creativity to market?);
3. ascertain what is needed to educate a *creative workforce* and wealth creators (in both formal and informal education);
4. experiment with prototypes of *content creation* and distribution, especially among non-specialist populations and socially disadvantaged groups; and

5. promote the inclusion of *creative innovation* into national innovation systems, where policy settings in many countries at the time were heavily skewed towards bioscience, ICTs and nanotechnology (why do countries need a creative economy?).

Thus our research began to focus on problems at the intersection of culture and economy. On the cultural side was creative talent, both 'expert' (artists) and population-wide (consumer co-creation); and on the economic side were innovation and growth in a knowledge-based economy.

When we got started, the very term 'creative industries' was contentious, imprecise and of dubious provenance, having originated in the opportunistic hot-house atmosphere of political competition and local boosterism, rather than from any discernible first principles, either economic or cultural.

POLICY

An existing science did not develop a theory of creative industries that might then be tested using formal hypotheses, experiments, fieldwork, data-analysis and the like. Instead, government departments sought to benefit from the 'new' or 'weightless' information- or knowledge-based economy,[2] from existing competitive advantage in certain industry sectors, especially in the UK, which boasted a large creative economy in the capital, and from seeking to redefine culture as an earning sector (growth) rather than a spending one (heritage, welfare).

All of this came together during the first New Labour government in the UK, when in 1998 culture minister Chris Smith sought to boost his portfolio's clout by yoking these 'heterogeneous ideas by violence together' (as if he were one of Dr Johnson's metaphysical poets), in order to get more support from the Treasury for culture.

The idea worked. You don't have to agree that it was a stroke of genius to admit that it was timely and productive and had far-reaching consequences, many unforeseen. Although the Treasury remained as flinty-hearted as ever, the 'creative industries' genie was unleashed from the knowledge-domain bottle, and Smith's department of Culture, Media and Sport (DCMS) gained first-mover advantage in defining the creative industries.

The important policy move here was to get culture away from the back door of the economy where it traditionally sat, tin cup in hand, crankily biting the hand that fed it, right around to the front and centre of innovation strategy, where suddenly it was revealed as a high-growth sector, outperforming other services (never mind sluggish manufacturing). It was dynamic and emergent, with multiplier effects on other sectors, a high rate of entrepreneurial initiative, and lots of start-ups, micro-businesses and sole traders, some of them – ageing rock stars – worth more than many large-scale companies.

FLAWS

One question did not even make it on to the agenda of a department whose main jobs were to spruik business and keep the arts lobby docile, and that was this: is it possible to have a 'creative economy' based on the creativity of the *whole population*, not just on existing artistic elites, professional designers and an 'expert pipeline' model of copyright-protected creativity?

There were other flaws in this policy initiative. First was the insistence that the creative industries, based as they are on individual talent, could prosper only in a world where *intellectual property* was strongly enforced. This left out of account the burgeoning world of consumer-created content and user-led innovation; and it forgot that creativity involves a lot of copying from past masters and contemporary competitors.

Second, it anchored the idea of the creative industries in the *analogue* era, where individual artists produced individual works either for public institutions or in a traditional marketplace dominated by single-platform firms (broadcasters, record labels, film studios, publishers, newspapers, fashion designers and so on). It missed out not only on the affordances of *digital* technologies, but more importantly on the *internet ethos* of 'knowledge shared is knowledge gained' and on the non-market or 'gift economy' aspect of social networking, crowd-sourcing and communities of affect.

Third was City of London hubris. After all, didn't the Brits have a world-beater in the shape of the City's financial services industry? And wasn't London a great creative capital in showbiz, publishing, media, broadcasting and cultural tourism as well? So what could be more appropriate than to model an ambition for the creative industries on the success of financial services, Britain's biggest export: all market, no regulation (this was when the light-touch FSA still seemed like a good idea); high levels of 'creativity' in product design (where the value-add is in the sophisticated inventiveness of products); high liquidity; high debt ('leverage'); and high compensation, all adding up to high-growth services. It all looked great as long as growth continued and credit was abundant. Surely the creative sector – like Pygmalion – could undergo a make-over and 'be like that'?[3]

The creative sector certainly produced some winner-take-all celebrities, like J.K. Rowling and Damien Hirst, and both creative and high-tech winners made it to the under-40s Rich Lists. But that did not translate into a creative economy more generally, nor did the creatives out-compete the bankers on the Hertfordshire-mansion-buying circuit. The richest person in Britain was still the Duke of Westminster (property), later overtaken by Lakshmi Mittal (steel) and Roman Abramovich (oil).

And in any case, the City of London was in for a shock, in September 2008, when the head office of Lehman Brothers in the US phoned to say, 'London, you're on your own.'[4]

A fourth weakness was its *nationalistic* bias. Everyone from the then UK finance minister Gordon Brown downwards believed that here was an example of 'competitive advantage' for the UK *as opposed to* other countries. All the talk was of how to *lead* the world, not *join* it. No one paused to wonder how countries might collaborate rather than compete in a globally networked system whose real motive force was located offshore.

Chris Smith's successor at the DCMS, Tessa Jowell, made speeches about how the UK could leave low-cost manufacturing to the Chinese and concentrate on high value-added creative goods and services. She forgot to mention that the Chinese themselves might have other ideas about that, and were listening carefully to foreign advisers (we were among them) who were telling them that they needed to shift from a low-cost 'made in China' economy to high-value 'created in China',[5] by growing their own creative sector, encouraging domestic consumption, and aspiring to turn around the 'creative trade deficit' where they imported more ideas from the West than they exported Chinese culture, media, branding and knowledge.

TECHNOLOGY

So while the creative industries initiative of the DCMS in the late 1990s was inspired, it was also self-serving and irrationally exuberant. The bubble obligingly burst in March 2000 with the dotcom crash, bringing the NASDAQ back from 5000 points to 1300, where it languished for years.

Although a good many ICT ventures lost their money, the digital media and the internet did not go away, any more than railroads or automobiles had after previous stock-market crashes in the 1850s and 1930s. Nor did the creative industries suffer as much as other sectors with exposure to ICTs. However, it was a chastening experience to see how precarious creative enterprise could be, among both (venture) capitalists and (artistic) workers, once culture had been redefined in market terms rather than in those of heritage or subsidy.

By this time, in fact, it had become clear (not to everyone; the arts have more than their fair share of resistance-to-change curmudgeons) that high-tech ICTs – fat pipes – were going to be crucial to the creative sector, not the death of it. European-style 'analogue' creative industries and cultural institutions did not amount to much without US-style digital technologies and market-based new-media platforms.

Here the European tradition of public culture and cultural institutions met the American tradition of individualism and the entrepreneurial ethos. Where 'Britart' artists might aspire to place their work in a museum (preferably the Tate Modern, Cool Britannia's latest tourist attraction), Californian computer geeks aspired to turn their string of code into a global corporation.

Was it possible to integrate these aspirations – intermingling artistry and entrepreneurship, individual talent and global scale, public culture and consumer demand, creativity and computing power?

In short, might digital technologies enable us to take creativity to population-wide participation and global scale? Did 'global media' have to imply 'monopoly control' by the usual suspects – Hollywood and international media moguls like Rupert Murdoch? Or could anyone get a look-in? If so, might smaller or emergent economies (say, Australia or China) benefit from technological advances and join in too?

Although 'big media' remained prominent as they migrated to the net, it was obvious from the start that online creativity could now also include a bottom-up, peer-to-peer element, since that's how the whole thing was invented in the first place. There was no reason in principle why such inventiveness had to be located in California.

GEOGRAPHY

Even so, California was a formidable competitor. Silicon Valley provided the model of a geographical *creative cluster*, where concentrating garage start-ups together seemed to make it easier for some of them to burgeon into global corporations in just a few years.

Richard Florida popped up to argue that the 'creative class' of knowledge professionals and ideas entrepreneurs was numerically small (only 150 million people worldwide), but it was disproportionately responsible for economic growth and creative innovation (which were increasingly the same thing). Its members included a lot more occupations than the traditional idea of creatives. There were the computer and mathematical geeks; architecture and engineering; life, physical and social-science occupations; education, training and library occupations; the arts, design, entertainment, sports and media.

These 'no collar' professionals, who also liked to live in an 'experience economy', were mobile and went where they liked, so you'd better make your city creative-friendly if you wanted to attract them. Florida's message struck a nerve with city planners worldwide, resulting in the surreal scene of mayors and bureaucrats puzzling over their 'creative class indexes' to see if they had enough students and gay people to make a viable creative city.

But could anywhere be a creative city? Even Florida didn't think so.

Eventually he settled on 40-odd 'mega-regions' as the crucibles of global creativity: 'The places that thrive today are those with the highest velocity of ideas, the highest density of talented and creative people, the highest rate of "urban metabolism"' (*The Atlantic*, March 2009).

Most of them remain in the USA, according to Florida, but worldwide you can also find them in Europe (Greater London; Am-Brus-Twerp) and Asia (Greater Tokyo, China's Shanghai–Beijing Corridor, and India's Bangalore–Mumbai area).

GROWTH

At this point, with the *convergence* of telecommunications, computer and media technologies, it was possible to imagine content creation as a globally distributed user-created system.[6] *Everyone* (with access to the web) could produce and publish their own media content, or share favourite stuff with their peers. Creative content converged with telecommunications. Instead of trying to make money out of unique items (this film, that painting), you could make it by promoting *creative traffic* among peers who made the creative content for themselves.

In other words, once the World Wide Web could handle video (by 2005), it was possible to imagine how something like BitTorrent or YouTube might replace broadcasting as the 'platform' for creative media – not one-to-many mass entertainment but 'many-to-many' messages (more like telecoms than media).

'Platform' is the wrong metaphor here, implying something stable upon which to build the castle where – as they said at the time – 'content is king'. The speed of change on the technical side continued to obey Moore's Law – growth was faster than exponential, doubling the extent of creative infrastructure, speed, connectivity, users, uses and content every couple of years. And of course much of the resultant content was shared, pirated, unpaid or amateur, making it very hard to erect a viable business plan over any new platform.

This produced further uncertainty and dynamism in creative enterprise. This year's hot new platform or 'killer app' was next year's landfill – thereby giving rise to a not altogether welcome new 'creative industry', that of processing e-waste, in which the Chinese town of Guiyu in Guandong is a world leader.[7] Intel's business plan assumed that the most successful product of the forthcoming financial year would be something that had not yet been invented. They maintained a research budget the same size as Australia's to make sure the invention was theirs.

Continuing growth was driven by technological innovation, by the extension of digital participation across an ever-wider population, and by the burgeoning uses to which all this capability could be put, both within businesses and in informal social networks.

One very interesting aspect of this growth was how it outpaced public policy settings. Although (as ever) crucial technological breakthroughs were part-funded by the defence industry, most of the energy came from non-government agencies, some set up for profit and many not.

When public policy did catch up – in the shape of the 1998 DCMS initiative – it focused exclusively on *economic* growth, as a sort of updated industry policy. It was not focused on the wider and more important question of *the growth of knowledge*.[8]

Thus, relatively little public investment was made in the *propagation* of digital take-up across populations, in education for digital literacy, or in support for creative development and organization (other than business services for creative firms).

In the nineteenth and early twentieth century, modernizing countries had invested vast public resources in achieving universal print-literacy, but no such effort was contemplated in relation to digital media. As far as the growth of knowledge via computers, telecommunications and media networks went, 'the people' were on their own.

SOCIAL NETWORKS

In short, the *growth of knowledge* was a problem not for the government but for the market – and that was government policy. If people wanted a creative economy benefiting from global digital technologies, and if individuals wanted to hook up with like-minded others worldwide, then they must 'do it themselves'.

The dotcom crash and the digital revolution emboldened some to say that the problem had been caused not by too much creativity but too little. Attention had been focused too narrowly on connectivity and the use of IT for internal business operations. What might the population at large like to do with this network? Enter DIY culture, Web 2.0 and the new global players, Google, Facebook, YouTube, Wikipedia and Flickr.

Here the true nature of the creative economy crystallized in a way that had not been clear till now. The creative industries were not the 'copyright' industry; they were not the 'arts' industry; they were not creative 'professions' (designers, media producers and so on); they were not the 'media industries'. The creative industries were characterized by something rather different: they were – and are – *social network markets*.[9]

Social network markets have two main peculiarities. The first is that people's choices are determined by the choices of others in the network. The second is that choices are status-based. Why are these characteristics peculiar?

1. Markets are supposed to be based on self-interested choice, which is assumed to be individualist and rationalist, not determined by the choices of others. In social network markets choice is externalist or system-based, produced by relationships, not reason – reason is the *outcome* of collective choices in a system of relations, not an *input*.
2. Choice is meant to satisfy wants or needs. But in social network markets it expresses status relations. Thus, the creative industries don't look very much like a neoclassical market. The choices of high-status celebrities will often be preferred, and those of low-status people avoided, creating a market in celebrity endorsement. Celebrity itself is not a product of, but an input into, such a market. Emma Watson can make Burberry cool again; but she had to be Emma Watson first.

'Entrepreneurial consumers' too can gain status by making admired choices (not just in high fashion but also in street fashion, such as Harajuku in Japan).[10] And because status is both relative and transient, the continuing process of making choices in social networks has an impact on status and thus on values (and further choices), both cultural and economic.

This is what Jason Potts calls 'choice under novelty' as opposed to choice under uncertainty or choice under risk, both of which have been studied in behavioural economics.[11] When faced with new knowledge, new connections or new ideas, people cannot reduce uncertainty by getting more information, precisely because what they're facing is new. Thus, suggests Potts, 'rational economic agents' – that's everyone – observe and learn from how others are making choices, and thus how to respond to the new. This is how they get into social network markets in the first place. Once there, observing and connecting with others, not least by random copying,[12] new possibilities open up, including other opportunities for 'consumer productivity' and co-creation.

Further, much of what constitutes social networks, and therefore the creative industries, is not market-based at all, at least not in the usual sense. This is because social networks exist prior to and outside of markets (among families, friends, neighbours, enemies and so on), and because they belong to the '*economy of attention*' as much as to the monetary economy.[13]

People place value on the *attention* they give and receive. This is an economy of *signals* as much as one of monetary values, which is why it needs a 'convergence' of cultural studies (semiotics, anthropology, media analysis) with economics to make sense of what's going on.[14] People may invest time, creativity and material resources in creating the right signals to attract more attention. They also value *paying* attention to favoured others. Fans, for instance, invest in the attention they offer to their idols. Attention may be 'paid' in many ways, not all of them monetized. Choice may just as easily end in a marriage or friendship as a sale.

CONSUMERS = PRODUCERS

The very concept of a consumer is irrelevant in social networks. Self-organizing networks of people, who are in it for the value of the relationship with others, are not really consuming anything. Quite the reverse. What's important is not what they buy, but what they make and how they signal, whether that's simply 'making sense' of stuff they like, or making contact with each other, or making their own creative content, from photos or text to competitive gaming strategies or open-source code. The erstwhile 'consumer' is now the focus and engine of the *productivity* of the system.

Social networks are not made of passive consumers waiting to be persuaded whether to buy the blue one or the white one, or push this button rather than that. They are strictly peer-to-peer, self-created and sustained, multi-nodal and mutually interconnected *networks*, not the end-point of a linear product pipeline. A *ballistic* strategy, where you 'target' this or that consumer profile and then 'bombard' it with well-aimed messages, is also wrong-footed.[15]

In a social network market you can't *make choices for* the consumer. The whole point is that users are doing the social networking for themselves. In essence this is a socio-cultural rather than an economic activity. What people are doing is about *their* status, and those they admire (or otherwise), including their own personal identity-forming activities and people in their own private circle.

CREATIVE DESTRUCTION

From this perspective – the perspective of the DIY user or 'productive consumer' – the status of organizations is not all that crucial. People interact with *content* and with *others*, not with *firms*. Maybe that's why people don't think of sharing as piracy – they don't see themselves as being in a proprietary environment.

It is easy to see that established distinctions between producer and consumer, public and private, property and piracy, expert and amateur, and agent and institution are undergoing a thorough process of Schumpeterian 'creative destruction'. This is 'remix culture' with a vengeance.[16]

In this context it is unwise even to hang on to 'the firm' as the obvious unit of agency or enterprise, since firms are by no means the only (certainly not the first) source of innovation in the online environment. Other forms of association, organization and institution have been established, from self-selecting networks of mutual interest to giant enterprises based on attention (Perez Hilton), the gift economy (Project Gutenberg), corporate branding (MIT's free courseware) or friendship (not just Facebook).

Many of these start out as amateur hobbies (genealogy), which subsequently prove robust enough to sustain both commercial and community forms of organization. In these kinds of networks, firms co-exist symbiotically with community networks. Very often the 'generative edge' of a new affordance is not motivated by profit, but unforeseen popularity may create a market which firms can stabilize.

In other words, there's an evolutionary process where the necessary variation and experimentation precede selection (firms), adoption (markets) and retention or extinction (competition). Thus the system as a whole is larger than the market aspect of it, and includes more kinds of enterprise than the firm, and more kinds of motivation than profit or price incentives.

So here's another peculiar thing. If, as government departments and industry analysts tend to do, you focus exclusively on the *producer* end of an industry, then you're likely to miss the creative industries altogether. They cannot be deciphered by looking at what *firms* do, only by looking at what *people* do, especially when they are interacting within very large-scale open complex systems.

But at the same time it's no good reducing 'people' to the status of an atomized and individualist 'self-contained globule of desire' (in Thorstein Veblen's words).[17] Individual identity is itself a social project, *produced in and by* the systems, institutions, networks and relationships in which they participate, all the way up from family and oral language to mass-mediated celebrity.

When you consider the 'identity' of global celebrities – let's say Paris Hilton – it is clear that this identity is a constantly produced work in progress, and that Ms Hilton is more like a global brand or firm than an individual. Of course, as a winner in the economy of attention she has amassed myriad more networked connections than 'ordinary' consumers, which means that in Barabasi's terms she's a 'hub' rather than a 'node'.[18] She's a spike and you're part of the 'long tail' on a 'power law' curve of attention connections in a 'scale free network'. Or, to put it another way, *everyone* is part of this interconnected network. Thus, despite individual difference between Paris Hilton and those who live 'the simple life', the identity of all is a product of connections in the same dynamic system.[19]

RADICAL RETHINK

To proceed on a business-as-usual business plan, where the script says that if firms target consumers then industry prospers, prematurely closes down the transformational potential of the times.

Instead of this, and precisely because we seem to have entered a topsy-turvy world where it is no longer clear what anything means, we need to take a good hard look at how open complex systems work, and what the role of creativity – not to mention industry – may be in that context.

As mentioned at the outset, this is exactly what the Centre of Excellence for Creative Industries and Innovation (CCI), funded by the Australian Research Council (ARC), has set out to do. As we've grappled with these issues over the years, it has become increasingly clear that we need a radical rethink about first principles.

As part of the work of the CCI, I have been engaged in a five-year 'Federation Fellowship', also funded by the ARC, to investigate 'the uses of multimedia'. My own research interest has been in what happens to previously popular media, especially broadcast television, in the digital and interactive era, and what *could* happen with population-wide digital literacy if you take the internet to be as significant an invention as printing was in the early modern period (which I do). My books *Television Truths* (Wiley Blackwell, 2008) and *The Uses of Digital Literacy* (University of Queensland Press, 2009) pursue these questions.

Printing with movable type was initially used (from the 1450s) for ecclesiastical and state purposes. Inadvertently, however, by the seventeenth century it had enabled the society-wide adoption of *realism* through three great textual systems, all of which required printed books and periodicals. These were *science*, *journalism* and *the novel*.

Given that history of unforeseen consequences and an out-of-all-proportion growth of knowledge resulting from the adoption of a new communications technology, what might be the consequences of interactive computer-based communications and the concomitant spread of *digital* literacy?

It still is too soon to tell, but it is obvious that we ought not to be thinking about the instrumental purposes of the opening players.

Lucy Montgomery has been one of three post-doctoral fellows working on the Federation Fellowship team, during which period she has written this book. The other post-doctoral fellows are Jean Burgess and John Banks. Jean Burgess works on 'vernacular creativity', and her co-authored book on *YouTube: Online Video and Participatory Culture* was published by Polity Press in 2009. John Banks has been studying the relationship between games companies and players, observing the 'community relations' aspect of games companies. Here the whole question of user-created content, and the distinctions between producer and consumer, expert and amateur, become life-or-death issues . . . for the games company. If its player community don't like a game, and the games developers won't listen, the company may founder. Banks's book *Making Co-creative Culture: Videogames and Social Network Markets* will be published in 2010–11.[20]

Our project benefited from contact with specialists in other fields, especially Centre fellows Jason Potts (evolutionary economics), Michael Keane (China's creative economy and policy) and Brian Fitzgerald (copyright and intellectual property law). During the course of our work, which also features decisive contributions from the sinologist-economist Carsten Herrmann-Pillath (Frankfurt School of Finance and Management), it has become increasingly clear that we can't just accept the definition of the creative industries at the going rate. There had to be a revaluation (see http://cultural-science.org).

HISTORY

We started this off by historicizing the very idea of the creative industries. It did not emerge from a definition, but from a situation, and therefore the idea as an 'artefact' is not scientific but historical. Thus, in *Creative Industries* (2005) and in Hartley (2009), I posited that the creative industries could already be seen as a dynamic, evolving concept even during the decade or so of its current usage.[21]

But before we get to that, a longer history needs to be acknowledged, going back to the Enlightenment (and thence to Classical antecedents). The public function of creative and cultural practices is constantly reinvented for each new era, so that the stage on which the creative industries debuted was already crowded with *dramatis personae*:

- From the *Enlightenment* notion of liberal arts and civic humanism (the public virtue of the gentleman) comes the idea of the creative arts as noble, civilizing, uplifting and aristocratic; '*noblesse oblige*'.
- From the nineteenth-century rise of the *nation state* came the idea of national culture and public arts, when European aristocrats turned their pictures over to national galleries and their palaces into museums.
- From *industrial culture*, especially in America, came the idea of popular arts – fiction, cinema, media – based in the marketplace, not the public institution, giving rise to the citizen-consumer as an amalgam of democratic values (freedom, public) *and* capitalist ones (comfort, private) – Walt Whitman *and* Wal-Mart.
- From *modernism and the artistic avant-garde* came the idea that creativity equals 'the new' and that anything produced at mass scale is a mere reproduction, thereby recasting aristocratic elitism into intellectual elitism, where the newest idea rules.
- From the *Frankfurt School* and anti-capitalist leftist academics came the idea of the *culture industries*, where media and state power were

coterminous and the state was controlled by capital; creative 'industries' were but a capitalist mouthpiece.

- *Regional policy* depoliticized the idea of the *cultural industries* and
  sought to attract them to set up in this or that country or city, resulting
  eventually in the idea of 'cultural capitals' and 'media capitals'.
- From the *information industry* came the idea of creativity as part of the
  'weightless', new or knowledge economy, adding high-value creative
  content to information infrastructure and connectivity, making creative
  'inputs' part of value-added services or intangibles.

All of these traditions were in place – and some of them in contention with
each other – when Chris Smith had his good idea that the creative industries
were an emergent economic sector. Now the question became: what kind of
economic sector was it?

EVOLUTION

Not surprisingly in such a fast-moving and turbulent context, the very idea of
what constitutes the creative industries is equally dynamic. Barely a decade
on, the concept has shown three distinct phases over its short life. Each phase
designates a different field of creative practice, each wider than and encom-
passing the one before. Thus you might model the creative industries idea as
a virus, spreading through a population via various hosts until everyone is
infected with it.

The phases are:

1. *Creative clusters (industry) – closed expert system*
   The first phase is the *industry* definition (DCMS), which I call 'creative
   clusters'. It is made of clusters of different 'industries' – advertising,
   architecture, publishing, software, performing arts, media production, art,
   design, fashion and so on – that together produce creative works or
   outputs. This is a 'provider-led' or supply-based definition. The sector is
   reckoned to be anywhere between 3 and 8 per cent of advanced
   economies, and claimed to be high-growth, with an economic multiplier
   effect.
   Looked at this way, the creative industries are nothing other than *firms*
   whose livelihood depends on creating intellectual property and protecting
   it with copyright, enforced against both commercial copying and
   consumer 'piracy'. It is thus modelled on the industrial-era closed expert
   pipeline (invent – patent/copyright – manufacture – distribute – sell),
   preferably with all of these functions controlled by the one firm.

2. *Creative services (economy) – hybrid system*

The second phase is the *services* definition, which I call 'creative services'. It is characterized by the provision of creative *inputs* by creative occupations and companies, most obviously where professional designers, producers, performers and writers add value to firms or agencies engaged in other activities, from mining or manufacturing to health, government and other public services.[22] By one estimate, creative services expand the creative industries by at least a third.[23] Again, the input is high value-add; indeed, it is thought to add value to the economy as a whole, boosting the profitability of otherwise static sectors.

It is this kind of creativity that transforms old-style services like transport into creative services like 'experience'-based tourism. Because this version of the creative industries is economy-wide and involves *occupations and agencies* other than firms, it may be regarded as a hybrid system, in which social networks play a role, but it remains focused on market-driven activity; it is 'demand-led' only in a b2b environment. Nevertheless here is where innovation policy can gain traction, encouraging firms of all kinds to collaborate with creative entrepreneurs and to innovate using creative inputs.

3. *Creative citizens (culture) – open innovation network*

The third phase is the *cultural* definition, which I call 'creative citizens'. Here is where creativity spills out of the economy, being an attribute of the population at large – the workforce, consumers, users and entrepreneurs, who become hard to distinguish from artists in how they go about pursuing an idea and creative reputation and a market for it. This is a user-led or demand-side definition. The expansion of the creative industries to cover everyone (at least in principle) allows the possibility that the energies of everyone in the system can be harnessed, adding the value of entire social networks and the individual agency of whole populations to the *growth of knowledge.*

Such a vastly expanded definition of creative agency is only 'thinkable' with complexity/network theory and the notion of open complex systems. It is most easily evident in computer-based social networks, but is not confined to the digital domain. Creative citizens are 'navigators' rather than 'consumers'; they may also act in concert as 'aggregators' to produce 'crowd-sourced' solutions to creative problems.

Following this line of thought it is easy to see that there's more to creativity than what is taught in art schools – or business schools. Creativity is generalized as a population-wide attribute, it requires social networks, and its 'product' is the growth of knowledge, sometimes within a market environment, sometimes not.

This is a radically democratic move, although it is far from universally adopted and its implications have barely begun to be worked through. But it is possible to identify the new value propositions associated with an evolved and expanded notion of population-wide creative industriousness in this formula:

Agents (both professional and amateur)
+ *Network* (both social and digital)
+ *Enterprise* (both market-based and other forms of purposeful association)
= *Creative value* (in a complex open system)
= *Growth of knowledge*

MORE MODELS

This 'social network markets' model of creative industries was first elaborated in an article by Jason Potts and others in 2008.[24] It continues to form the basis of the work we're doing at the CCI, and it is given further evidential substance here in Lucy Montgomery's book.

A different model, seeking to account for the creative economy along a different line of thought, was published by Potts and Cunningham at around the same time.[25] Instead of showing the phases by means of which creativity ripples out from industry, via economy, to culture (and back again), this approach sought to show that differing economic theories and approaches brought a different model of the creative industries into view. The four models are:

1. *welfare* (market failure model of culture), requiring a *negative* policy of *welfare subsidy*;
2. *competition* (nothing special here), requiring a *neutral* or standard *industry* policy;
3. *growth* (creative industries as a dynamic sector – DCMS-style), requiring a *positive* investment and *growth* policy;
4. *innovation* (creative industries as general dynamic of change), requiring an *evolutionary* policy of *innovation*.

All of this modelling is characterized by one major concern – the role of creativity in change, both cultural and economic, and its location as a property of agency in open complex systems ('networks').[26]

The creative industries were looking less like a small but sexy sector in affluent economies,[27] and more like a general *social technology* for enabling change.

It may even be argued that the 'creative industries' are the empirical form taken by innovation in advanced knowledge-based economies. This would place *creative innovation* on a par with other enabling social technologies like the law, science and markets. The creative industries may be regarded as the social technology of distributed innovation in the era of knowledge-based complex systems.[28]

Whichever way you looked at it, we were heading for an *evolutionary* approach to culture in general. Modelling not just the economy but also culture as evolutionary, and seeing creativity as part of the general process of innovation and adaptation to change, has led us towards a new kind of intellectual enterprise that goes under the heading of 'cultural science'.

THROWING BRICKS

But just before we get to that, it is worth bringing geography back into the picture. If the creative industries can be seen in terms of 'social network markets', then any industry has to go through a 'creative industries' phase at some point, because the creative industries 'involve the creation and maintenance of social networks and the generation of value through production and consumption of network-valorized choices in these networks'.[29]

Thus there is a development aspect to creative industries thinking, because, if adaptation to change, access to technologically enabled digital networks, and the production and consumption of network-valorized choices among a creative population do constitute the creative industries, then developing and emergent economies need them more than anyone.

Furthermore, without being encumbered by industrial-era investment in smoke-stack industries and rust-belt regions, emergent countries may aspire to become 'leapfrog economies', using the creative industries as a social technology of modernization, global engagement and urban development.[30] This stimulates their SME and NGO sectors and promotes the development of indigenous micro-business.

It also helps to develop the most abundant resource available to developing countries, *creative human capital*, especially with their predominantly young populations. For young people especially, creative expression is itself an attractant to enterprise. In developing countries, a creative economy is also a powerful tool for promoting and valorizing diversity, of both population and cultural expression, both traditional and modern. Thus creative industries are the generative edge of innovation among the billion or so young people worldwide who are now moving through their teenage years towards full economic productivity.

That is why it is just as important to consider countries like China as the USA or UK – indeed the model of creative industries inherited from the latter may be disastrous for emergent economies, based as it is on analogue technologies like painting or the record industry.

Thus whatever model of the creative industries is adopted, it needs to take account of the 'creative destruction' that may imminently be wrought on the global economic and cultural scene by the BRICKS countries – Brazil, Russia, India, China, Korea, South Africa – and, one may add, Indonesia (and many others – this being the point).

Lucy Montgomery's book is one of the first to take the idea of the creative industries and work through it in the context of an emergent economy – in this case the strategically crucial one of China.

COPYRIGHT CONFLICT

In a context such as this, it is unwise to carry forward a definition of the creative industries that is based on record-label and Hollywood notions of copyright. Militant enforcement of owners' IP rights and anti-networking stratagems like DRM are predictable responses by existing investors, but that doesn't make them good policy for new ones.[31]

But at the same time it is obvious that commercial value requires that you have something to sell, and if that something is an idea there has to be a way to monetize it. Hence there is no escaping the fact that copyright and intellectual property are *the* point of tension for contemporary cultural, creative and commercial conflicts for the foreseeable future.

Here is where old and emergent enterprises clash, and, if the overall 'growth of knowledge' is the ultimate object of study, then the real and contentious questions of who owns that knowledge, how it can be shared, through what media of distribution, and accessible by whom (and on what terms) are very pressing.

Yet neither economics nor cultural studies have thought as carefully as they should about the problem of copyright. It's not an old-style 'Left v Right' struggle between libertarian progressives and control-culture reactionaries (though it does look like that sometimes). It is a problem of how to coordinate and organize innovation, dynamism and change in an existing complex system without snuffing out either the system or the change.

CULTURAL SCIENCE

And so it is more important than ever to get a clear idea of what's going on. Within the flux of change and differences of approach, opinion and purpose, it is never going to be possible to derive an analytical understanding from mere observation of the immediate to-and-fro of policy, debate, promotion and critique. However, it is equally clear that continuing to approach the creative industries using existing templates – whether by means of business plans, public policy or academic disciplines – will miss much of what has been described above.

This is what has led us to a work in progress that we're calling 'cultural science'. It is an evolutionary approach to both culture and the economy, combining evolutionary economics, cultural studies, and complexity or network studies. It seeks to investigate the growth of knowledge. It sees cultural and biological systems as co-evolutionary, and sees culture as a complex open adaptive system.

An immediate problem is that the fields closest to the study of creativity, including media studies, cultural studies, area studies (that is, the cultures and languages of particular countries) and various branches of the social sciences, have been among the very fields most resistant to 'taking the evolutionary turn' in relation to their object of study. Furthermore, the traditional 'two cultures' distinction between science and the humanities, and the lack of numeracy among many of the latter serve to make empirical studies of complex evolving systems hard to attempt.

There are equally gaping holes in the knowledge, skills and aptitudes of those coming to creativity and culture from economics or the sciences. They can do circulation, but they're not so astute about meaning. On both sides there is a tendency to adopt a 'heads down' attitude to concentrate on the micro-scale of local and familiar problems, leaving the macro-scale of systemic coordination, hierarchy, growth and interconnection with other systems out of the analytic picture.

Just because the phenomena under observation are both dynamic and multi-valent, it is therefore necessary to move from single-discipline research to problem-solving research, from solo hyper-specialization to team-based collaboration, and from national silos to international research networks. Most importantly, it is necessary to develop a coherent conceptual and theoretical framework through which to advance from an observational to an analytical approach.

Since embarking on this adventure, those of us pursuing a cultural science approach have become more firmly convinced that this is worth pursuing by finding that – under various banners – plenty of others are pursuing it too. There are well-established programmes of research into cultural and biological co-evolution in anthropology, language, neuroscience, economics and even social science.[32] More recently, evolutionary studies of stories, art and technology have been published.[33] The use of complexity studies and game theory to model and analyse social networks (both analogue and digital) is well advanced.[34] Cultural science is but one strand in this general current, and very much in its infancy.

However, it is worth the effort not only to find out how the growth of knowledge works, but also for more immediate gains. Those trained in the humanities have found it hard to make an impression on public policy formation or engagement with business, especially in relation to innovation. It is hard to persuade policymakers and business strategists to take creativity seriously without systematic and numerate evidential data to back up any claims.

Thus, while many agree that it is vital to add the cultural and human sciences to national R&D investment, to add creativity to science, technology, engineering and medicine (STEM) as an integral part of the innovation system, and to foster creative ideas as well as going for the technical fix, none of this will happen if those who are interested in culture and creativity can't speak the same language as those who are interested in the growth of both the economy and the knowledge base.

In the meantime, opportunities are being wasted, by government, business and cultural experts alike, to make better use of our growing understanding of the interfaces between cultural and economic values, between ideas and markets, between users and technologies, between elite expert systems and consumer populations, and between emergent and mature national systems.

READ ON

This is the interdisciplinary context in which Lucy Montgomery's book fits. She has pursued practical questions with this context in mind, while seeking as far as is currently possible to adopt an approach that aligns with cultural science as it unfolds. The result is a forward-thinking book that looks at China in a new way and comes up with some new questions.

China's Creative Industries provides a coherent argument with real drive and purpose. It advances our theoretical and conceptual understanding of the creative industries, of intellectual property, and why these things need to be thought about differently. It shows why China is important to the overall situation and not just a regional application. I really like the build-up of concepts, from 'entrepreneurial governmentality' and the 'entrepreneurial consumer' to a reworked notion of intellectual property that applies not only to China.

This book adds substance to our team's earlier conceptual work, joining the agenda-setting publications mentioned above as part of a coherent intellectual programme.

Together with further work by Potts, Banks, Burgess, Herrmann-Pillath and others, Lucy Montgomery's book also adds substance to our 'cultural science' claims. We have moved beyond the merely observational (I went to China/the games industry/fashion and saw this activity, spoke to that individual) to the properly analytical.

Whether you're interested in economic, cultural, technological, Chinese or copyright issues, I commend *China's Creative Industries: Copyright, Social Network Markets and the Business of Culture in a Digital Age* to your attention. As Montgomery demonstrates, the problem of indeterminacy in how we understand the creative industries that I have tried to elaborate above – the question *'Whose* creative industries?' – does have one emergent but clear answer: *China's* creative industries.

John Hartley, AM
Brisbane 2010

NOTES

1. John Hartley, AM, is Australian Research Council (ARC) Federation Fellow and Research Director of the ARC Centre of Excellence for Creative Industries and Innovation at Queensland University of Technology (QUT). Prior to that he was foundation dean of the Creative Industries Faculty at QUT and foundation head of the School of Journalism, Media and Cultural Studies at Cardiff University. The author of 20 books and many articles in the fields of media and creative industries, he is a Fellow of the Australian Academy of the Humanities and editor of the *International Journal of Cultural Studies.*
2. C. Leadbeater (1999), *Living on Thin Air: The New Economy*, London: Viking.
3. See www.reelclassics.com/Musicals/Fairlady/lyrics/fairlady-whycantawoman.htm.
4. See: http://everythingneednotfit.blogspot.com/2008/10/betrayal-of-london-unheard-in-new-york.html.
5. M. Keane and J. Hartley (eds) (2006), *International Journal of Cultural Studies*, 9 (3) (September), special issue on *Creative Industries and Innovation in China*; M. Keane (2007), *Created in China: The Great New Leap Forward*, London: Routledge.
6. H. Jenkins (2006), *Convergence Culture*, New York: NYU Press.
7. See 'Electronic Waste' and 'Electronic Waste in Guiyu' (Wikipedia); and www.china-pix.com/multimedia/guiyu/.
8. B. Loasby (1999), *Knowledge, Institutions and Evolution in Economics*, London: Routledge; and see J.S. Metcalfe and R. Ramlogan (2005), 'Limits to the Economy of Knowledge and Knowledge of the Economy', *Futures*, 37 (7), 655–74.
9. J. Potts, S. Cunningham, J. Hartley and P. Ormerod (2008), 'Social Network Markets: A New Definition of the Creative Industries', *Journal of Cultural Economics*, 32 (3), 167–85.
10. J. Hartley and L. Montgomery (2009) 'Fashion as Consumer Entrepreneurship: Emergent Risk Culture, Social Network Markets, and the Launch of *Vogue* in China', *Chinese Journal of Communication*, special issue: *China: Internationalising the Creative Industries*, 2 (1), 61–76. And see www.japaneselifestyle.com.au/tokyo/harajuku_fashion.htm.
11. J. Potts (forthcoming), 'Can Behavioural Biases Explain Innovation Failures? Toward a Behavioural Innovation Economics', *Prometheus*.
12. See R.A. Bentley (2009), 'Fashion versus Reason in the Creative Industries', in M. O'Brien and S. Shennan (eds), *Innovation in Cultural Systems: Contributions from Evolutionary Anthropology*, Boston, MA: MIT Press, pp. 121–6.
13. Richard A. Lanham (2006), *The Economy of Attention*, Chicago: Chicago University Press.
14. C. Herrmann-Pillath (2010), *The Economics of Creativity and Identity: A Cultural Science Approach*, St Lucia: University of Queensland Press.
15. Hence the change in marketing ideologies, towards 'viral', 'groundswell' and 'herd' conceptualizations of marketing communication. See C. Li and J. Bernoff (2008), *Groundswell: Winning in a World Transformed by Social Technologies*, Boston, MA:

Harvard Business School Press; and M. Earls (2007), *Herd: How to Change Mass Behaviour by Harnessing Our True Nature*, London: John Wiley.
16. L. Lessig (2008), *Remix: Making Art and Commerce Thrive in the Hybrid Economy*, London: Bloomsbury, www.bloomsburyacademic. com/remix.htm.
17. T. Veblen (1898), 'Why Is Economics Not an Evolutionary Science?', *Quarterly Journal of Economics*, 12, http://socserv.mcmaster.ca/econ/ugcm/3ll3/veblen/econevol.txt.
18. A.-L. Barabasi (2003), *Linked: How Everything Is Connected to Everything Else and What It Means*, New York: Plume.
19. Paris Hilton's reality TV series *The Simple Life* (2003–07) won her the 'Innovator Award' at the 2009 Fox Reality Awards.
20. See also J. Banks and J. Potts (2010), 'Co-creating Games: A Co-evolutionary Analysis', *New Media and Society*, 12 (2), 253–70.
21. See J. Hartley (2005), 'Creative Industries', in J. Hartley (ed.), *Creative Industries*, Malden, MA and Oxford: Wiley-Blackwell, pp. 1–40; and J. Hartley (2009), 'From the Consciousness Industry to Creative Industries: Consumer-Created Content, Social Network Markets and the Growth of Knowledge', in J. Holt and A. Perren (eds), *Media Industries: History, Theory and Methods*, Malden, MA and Oxford: Wiley-Blackwell, pp. 231–44.
22. The first proponent of this view of the creative economy was John Howkins: see J. Howkins (2001), *The Creative Economy: How People Make Money from Ideas*. London: Penguin.
23. See P. Higgs, S. Cunningham and H. Bakhshi (2008), *Beyond the Creative Industries: Mapping the Creative Economy in the United Kingdom*, London: NESTA, www.nesta.org.uk/library/documents/beyond-creative-industries-report.pdf.
24. J. Potts, S. Cunningham, J. Hartley and P. Ormerod (2008), 'Social Network Markets: A New Definition of the Creative Industries', *Journal of Cultural Economics*, 32 (3), 167–85.
25. J. Potts and S. Cunningham (2008), 'Four Models of the Creative Industries', *Cultural Science Publications*, http://cultural-science.org/FeastPapers2008/StuartCunningham Bp.pdf.
26. See J. Potts (2008), 'Creative Industries and Cultural Science: A Definitional Odyssey', *Cultural Science Journal*, 1 (1), http://cultural-science.org/journal/index.php/culturalscience/article/view/6/15. Potts found 17 models!
27. E. Currid (2008), *The Warhol Economy: How Fashion, Art, and Music Drive New York City*, new edition, Princeton, NJ: Princeton University Press.
28. J. Hartley (2009), 'From the Consciousness Industry to Creative Industries: Consumer-Created Content, Social Network Markets and the Growth of Knowledge', in J. Holt and A. Perren (eds), *Media Industries: History, Theory and Methods*, Malden, MA and Oxford: Wiley-Blackwell, pp. 231–44.
29. J. Potts (2008), 'Creative Industries and Cultural Science: A Definitional Odyssey', *Cultural Science Journal*, 1 (1), http://cultural-science.org/journal/index.php/culturalscience/article/view/6/15, Definition no. 12. See also J. Potts (2009), 'Do Developing Economies Require Creative Industries? Some Old Theory about New China', *Chinese Journal of Communication*, 2 (1), 92–108.
30. See UNCTAD (2008), *Creative Economy Report 2008: The Challenge of Assessing the Creative Economy towards Informed Policy-Making*, www.unctad.org/creative-economy.
31. J. Potts and L. Montgomery (2009), 'Does Weaker Copyright Mean Stronger Creative Industries? Some Lessons from China', *Creative Industries Journal*, 1 (3), 245–61.
32. For instance the Centre for the Coevolution of Biology and Culture at Durham University (www.dur.ac.uk/ccbc); and see P. Richerson and R. Boyd (2005), *Not by Genes Alone: How Culture Transformed Human Evolution*, Chicago: Chicago University Press; A. Mesoudi (2007), 'A Darwinian Theory of Cultural Evolution Can Promote an Evolutionary Synthesis for the Social Sciences', *Biological Theory*, 2 (3), 263–75; J. Hurford (2007), *The Origins of Meaning*, Oxford: Oxford University Press; P. MacNeilage (2008), *The Origin of Speech*, Oxford: Oxford University Press; D. Bickerton (2009), *Adam's Tongue: How Humans Made Language, How Language Made Humans*, New York: Hill & Wang; W.G. Runciman (2009), *The Theory of Cultural and Social Selection*, Cambridge: Cambridge University Press.

33. See Y. Lotman (2009), *Culture and Explosion*, Berlin: Mouton de Gruyter; B. Boyd (2009), *On the Origin of Stories*, Cambridge, MA: Harvard University Press; D. Dutton (2009), *The Art Instinct: Beauty, Pleasure and Human Evolution*, London: Bloomsbury; B. Arthur (2009), *The Nature of Technology: What It Is and How It Evolves*, New York: Free Press.
34. Most obviously in the work of the Santa Fe Institute; see also E. Beinhocker (2006), *The Origin of Wealth: Evolution, Complexity, and the Radical Remaking of Economics*, Boston, MA: Harvard Business School Press; K. Sawyer (2005), *Social Emergence: Societies as Complex Systems*, Cambridge: Cambridge University Press; R.A. Bentley and P. Ormerod (2010), 'Agents, Intelligence, and Social Atoms', in M. Collard and E. Slingerland (eds), *Integrating Science and the Humanities*, Oxford: Oxford University Press; P. Ormerod (2007), *Why Most Things Fail: Evolution, Extinction and Economics*, London: Wiley.

Acknowledgements

If writing this book has taught me anything, it is that the life of a research academic is a privileged one. One of the greatest privileges is the opportunity to work with thinkers and innovators both within universities and in industry. Many of the people who have helped me to make this book a reality have been working in China's creative industries. Others have been writing about creative industries, digital technologies or the role of intellectual property as practitioners, teachers and students. All have displayed remarkable patience and kindness.

I am deeply indebted to my colleagues at Queensland University of Technology, particularly John Hartley, who provided me with the opportunity to pursue this research and the time and encouragement necessary to turn it into a book: the research on which this book is based was carried out while I was a post-doctoral research fellow on John Hartley's Federation Fellowship programme, funded by the Australian Research Council and supported by Queensland University of Technology between 2007 and 2010; Jason Potts, for his willingness to collaborate and to provide new perspectives on the questions I have been tackling; Michael Keane, who helped to supervise the Ph.D. that got this project started; and Brian Fitzgerald, Stuart Cunningham, Terry Flew, Christina Spurgeon and Stephi Hemelryk Donald, all of whom I have worked with on China-related projects.

I would like to thank the people who so generously allowed me to interview them for my research. Without the good will of the film, music and fashion industry professionals, lawyers and judges who took time to answer my questions and share their perspectives this book really would not have been possible.

I am especially grateful to the research assistants who helped me during my fieldwork: Udom Tangjettanaporn, Helen Gao, Vincent Weifeng Ni and Jingwei Li. Some of the greatest fun I have had as a researcher has been a result of staying with colleagues and friends during fieldwork trips: Bao Jiannu, Jenny Spark and Sen Lee; I would also like to thank the people who have provided me with encouragement and support throughout my research journey, particularly Vikki Katz, Helen Klaebe, Marcus Foth, Shaun Chang and my fellow post-docs Jean Burgess and John Banks.

This book has largely been written while I have been based at City

University London, and I would like to thank my hosts here for their inspiration and support, particularly Malcolm Gillies. Also in London, I would thank Birgitte Andersen and the Birkbeck Centre for Innovation Management, as well as the staff and students at the University of Westminster's China Media Centre.

Last, and by no means least, thank you to my family. I am especially grateful to my mother, Brenda Ford, who has been reading my drafts and providing feedback and encouragement for as long as I can remember. And thank you to Ben, for his patience and sense of humour.

Abbreviations

BFA	Beijing Film Academy
CC	Creative Commons
CCP	Chinese Communist Party
CEPA	Closer Economic Partnership Arrangement
CMCC	China Mobile Communications Corporation
CNNIC	China Network Information Centre
CPC	Chinese Communist Party
CRBTs	caller ring-back tones
DCMS	Department of Culture, Media and Sport, UK
FIPP	International Federation of the Periodical Press
GDP	gross domestic product
GPL	General Public License
HKTDC	Hong Kong Trade Development Commission
IFPI	International Federation of Phonographic Industries
IP	intellectual property
IPR	intellectual property rights
MPAA	Motion Picture Association of America
OECD	Organisation for Economic Co-operation and Development
PRC	People's Republic of China
RIAA	Recording Industry Association of America
SARFT	State Administration of Radio, Film and Television
TRIPs	Agreement on Trade-Related Aspects of Intellectual Property Rights
UNESCO	United Nations Educational, Scientific and Cultural Organization
USTR	United States Trade Representative
WIPO	World Intellectual Property Organization

1. Introduction: from governance to entrepreneurial consumers

This book is about content, technology, globalization and the challenges faced by creative and cultural businesses in a digital age. It deals in part with the tension between state-generated frameworks for managing creative industries and the changing mental technology that citizens carry with them in a rapidly evolving digital world. Most importantly, it examines the shift from state-controlled cultural production and consumption towards the emergence of 'entrepreneurial consumers' in China and considers what this might mean for the way that innovation, knowledge production and sources of growth are understood in the creative and cultural industries in the twenty-first century.

In the People's Republic of China (PRC), social, economic, political and technological transformations are having a powerful impact on the context within which creative and cultural activities occur. Although China remains a single-party state, the shift towards a market-driven economic system means that it is no longer possible for the government to dominate popular culture with the level of effectiveness witnessed under Mao. Economic reform has generated new space for market-driven activity, legal reforms have created new classes of property rights in creative and cultural works and brand identities, and digital technologies are transforming processes of production, distribution and delivery.

Systems that were once effective at controlling the ways in which popular culture was made and used are being challenged by an explosion in unauthorized distribution channels, as well as by the growing capacity of consumers to make their own content and to use existing material for new purposes. A market-driven economy, increased availability of information about cultural products and trends, new technologies and growing disposable incomes are all providing China's citizens with opportunities to understand consumption as a game with social costs and benefits: money, risk, status, fun and power. In navigating this new landscape, consumers are developing the skills and habits of *entrepreneurs* – with profound consequences for creative and cultural businesses.

In this book the shift away from centralized control of creativity and cultural production by the state towards much more dynamic models of creative consumption is traced through three of China's newly defined creative

1

industries: film, music and fashion. Film, music and fashion have been chosen because there are important differences in the ways in which state, corporate and consumer power are interacting in each of these industries. Businesses in all three areas are developing unique strategies for navigating changes in government regulation, as well as new commercial opportunities associated with cultural and economic reform. The impact of new technologies on business models and the extent to which consumers are able to influence creative production and decide how products are used and experienced also varies across these three sectors.

Revolutionary songs, operas and films have all been considered central to the Communist Party of China's (CPC) efforts to win the hearts and minds of the people, to imbue China's population with revolutionary fervour, and to convey political, ideological and moral messages to society (Kraus 2004). Fashion, too, has served political and ideological ends under the CPC: the body is a powerful site for disciplining the mind and the outward display of support for and conformity with the ideological goals of the state (Hooper 1994: 164). But, while film, music and fashion have all been associated with the exercise of state power over individuals at various times since 1949, the impact of economic reform and technological change on the growing business of culture in each of these industries has not been uniform.

I suggest that the film industry might be understood as an example of continuing *state* agency in the realm of cultural *production*. It must be said straight away that the reality of consumption in a country where unauthorized copying and distribution networks are so well developed is a different matter. By contrast, the role of *entrepreneurial commercial classes* is becoming increasingly obvious in the development of profit-focused elements of China's *music* industry, particularly in relation to mobile distribution opportunities. Finally, *fashion* and the fashion industry in China rely most heavily on the *active involvement of consumers in the productive process*. The status-driven dynamics of fashion consumption and the fashion system's ability to accommodate consumer co-creation and entrepreneurship are making it possible for China's fashion industry to develop in the absence of a strong intellectual property system. As China is becoming more deeply integrated into a global fashion landscape, the business models that are being taken up by China's fashion industry are taking on the characteristics of those seen elsewhere in the world, including consumer entrepreneurship as a driver of innovation.

Since the Open Door Policy of the early 1980s, the government's efforts to control the distribution and consumption of films have been seriously undermined by unauthorized distribution of cinematic content. At first this occurred mainly via VHS cassettes, which were copied and circulated between networks of friends and associates. Throughout the 1990s and the first half of the 2000s, increasing integration into international communities of trade, growing disposable incomes and the growing availability of new technologies for copying, distribution and consumption were associated with the growth of highly sophisticated networks for the sale of illegally copied cinematic works, prompting international calls for the Chinese government to clamp down on film 'piracy' and to increase access to the Chinese market for international copyright owners.

By early 2009 it was becoming harder to find shops selling 'pirated' DVDs on the streets of Beijing and Shanghai. To a casual observer it might appear that international pressure for better enforcement of copyright is beginning to have an impact on local practices. However, the declining visibility of 'pirated' DVDs is coinciding with technological changes that provide cheaper, more convenient means of accessing content for many urban consumers. By June 2009 China was home to 338 million internet users, 320 million of whom had broadband access (CNNIC 2009), and low levels of copyright enforcement online are making it possible for ever-increasing numbers of people either to stream or to download a very wide range of content through search engine links and dedicated film streaming sites, without leaving the comfort of their desks.

But, in spite of the rapid expansion of unauthorized distribution networks that has occurred in China over the past 30 years, the commercially focused film industry continues to be dominated by censorship policies and other government-administered structures of control. High costs associated with making commercially competitive films, the continued existence of pre- and post-production film censorship, and government control over cinema-based exhibition are preventing the film industry from breaking away from state influence. Cultural policies remain a defining feature and are limiting the spaces in which commercially focused film industry entrepreneurs can operate.

In certain areas of the music industry, on the other hand, lower production costs and less restrictive cultural policies are allowing greater space for a commercially focused entrepreneurial class to operate. Some Chinese entrepreneurs have been quick to understand music's potential to generate financial returns, as well as the commercial power associated with controlled distribution channels, such as mobile phone networks. Overall, there appears to be greater opportunity for an entrepreneurial class to have an impact on the music industry's development than in the case of the film industry.

The groups that are emerging as most powerful in terms of their ability to act as gatekeepers to lucrative distribution networks and commercial opportunities in China appear not to be the major record labels that so successfully established dominant positions within the commercial music industry in other markets. Rather, mobile phone operators and, to a certain extent, search engines are playing a powerful role in this emerging commercial landscape, challenging common Western notions of the way in which a recorded music industry functions.

In Western markets, common recording industry business models rely on communities of hopeful amateurs as an important source of 'talent'. Entrepreneurs and entrepreneurial firms (record labels) invest in development, production, promotion and distribution of the works of carefully selected artists, in return for the assignment of intellectual property and contractual rights. The assignment of these rights allows labels to recoup costs and generate profits from the sale of physical music products, as well as licensing on works in their accumulated repertoire. While new technologies for distribution and consumption of music are challenging this business model, even in the markets in which it has been most firmly established, such as the United States and Western Europe, labels continue to play an important role as brokers of new talent and coordinators of supply.

However, in China record labels have struggled to construct a similar role for themselves. Very high rates of unauthorized copying and distribution of physical media and difficulties associated with enforcing intellectual property rights online have prevented a record-label-centred model from taking hold. It appears that mobile operators and search engines, rather than labels, are dominant forces in the distribution of music in China, at least for now. But the role of music industry executives has not disappeared entirely, and the structure of the music industry, as an emerging area of economic activity, remains in flux. Although copyright is difficult to enforce, it is beginning to play a role in the negotiation of distribution rights. Furthermore, Chinese record labels and executives play an additional role to their counterparts in the West: negotiating government content regulations and ensuring that music and public performance comply with censorship policies.

In a market-driven economy that provides access to an unprecedented array of content, technology, products, services and experiences, the impact of consumer decisions on cultural production has increased. In some ways, this is a predictable outcome of market reforms in the cultural sector. However, transformative technologies have added new elements to the well-rehearsed story of supply and demand. The internet, high-powered personal computers, mobile devices and an increasingly lucrative services sector are changing relationships between 'producers' and 'consumers' of cultural products. It is becoming harder and more expensive to maintain control over distribution and reuse. Instant communication is increasing the power of social networks, allowing 'consumers' to apply creativity and skills to generate content and value, and to draw on information provided by trusted friends or communities in order to supplement that which is supplied by either the state or commercially interested advertisers.

Many of China's consumers are experimenting with different ways of interacting with creative works. It has become possible to 'play' with content in ever more personal ways, to use it to connect with others, to create an image of oneself, to strengthen social connections or to build social capital. Music can be 'displayed' to others as a mobile ring tone, personalized ring-back tone or web-page wallpaper. It can also be composed, mashed or remixed, and then shared with friends and online communities, who might enjoy it, pass it on or change it. Digital cameras, mobile phones and photo-sharing sites on the internet make it possible for people to tell stories about themselves and their world in new ways and are encouraging innovative forms of creative experimentation, collaboration and activism.

Technological developments that are making new kinds of creativity, interaction and distribution possible do not mean that existing structures for the production and distribution of creative and cultural works no longer have a role to play. However, disruptive technologies that make it possible for consumers in China and elsewhere in the world to access, use and distribute content and information in new ways raise a number of important questions about how value is generated in the creative industries. Do uncontrolled copying and peer-to-peer distribution prevent commercial development in the creative industries? Does an inability to enforce intellectual property rights remove incentives for investment in creativity and innovation? If so, is a strong intellectual property system a prerequisite for the growth of the creative industries? How might understanding this process better inform approaches to the protection of intellectual property in the context of digital technologies and global content flows, where controlling copying and distribution is more difficult than ever before?

One area that might provide some insight into changing dynamics of production, distribution and consumption in a digital age is fashion. Rather than being threatened by copying – of styles, looks and designs – innovation in the fashion industry is driven forward by these activities as consumers search for status and novelty and designers and retailers seek out new ways to provide it. Because status and identity are driving forces in the fashion industry, consumers are active participants in processes that create commercial value. The choices of lead consumers help to popularize new trends, information about choices is shared through social networks ('I love your hat – where did you get it?') and innovations made by consumers in the form of street fashion help to provide inspiration for design professionals.

Creative remixing is integral to processes of both fashion production and consumption, occurring when consumers attempt to imitate a 'look' or chose a flattering ensemble, or when designers scour vintage shops, art exhibitions or fashion magazines in search of inspiration for the next season's collection. The fashion system's dependence on creative remixing means that high levels of semiotic literacy are rewarded among both fashion professionals and consumers, who must understand the symbolic value of the elements that they are recombining in order to minimize the risk involved in their choices and to maximize the chances that their innovations will be admired and convey a desired message.

Fashion involves risks and benefits: rewarding consumers who become literate in the complex language and rules of the system and punishing those who deviate too far from accepted norms or make the 'wrong' choices. It encourages consumers to engage in thoughtful experimentation and to critically consider new trends in the context of their own tastes, preferences, values and desired associations. Because fashion provides consumers with opportunities for individual creativity, self-expression and innovation, the fashion industry is one in which 'consumers' are anything but passive. Rather, the fashion system demands that they are actively involved in a process of productive consumption.

Table 1.1   The value chain of three different creative industries in China, showing the shift from top-down control to bottom-up productivity

	Film	Music	Fashion
Production	High capital requirements.	Lower capital requirements. Diverse sources of production.	Distributed creative production. Consumer involvement in creative iteration is encouraged.
Distribution	State control of access to commercially significant distribution channels.	State dominance over commercially significant distribution channels.	Distribution is managed by the market.
Regulation	Commercial opportunities heavily dependent on spaces defined by the state. Pre- and post-production censorship.	Distribution licensing at the discretion of authorized publishers. Licensing for large-scale public performances. Centralized censorship processes for international artists.	Policies focused on creating an environment conducive to the growth of a fashion industry, rather than direct control.
Consumption	Passive.*	Interactive.	Participatory, entrepreneurial, productive.

* 'Passive' = semiotic use ('passive' does not imply agentless behaviour in this context).

In contrast to China's film industry, where commercially motivated production is dominated by organized professionals with access to capital, specialized equipment and state-controlled distribution networks, fashion involves much higher levels of distributed creativity and consumer engagement in productive processes. Consumers of fashion are encouraged to interpret, combine and apply the products and techniques of fashion professionals in the context of their own lives and aspirations. Success in the fashion game involves critical participation, rather than becoming a docile 'victim' of the latest trend. Experimentation and remixing, within the boundaries of 'taste', are admired.

Concepts of 'taste' are influenced not only by the fashion media, but also by a complex visual world that in twenty-first-century China includes newspapers, celebrity reporting (in magazines and online), film and television, as well as by the choices of friends, colleagues and associates. The role of honorific values and the absence of direct state intervention in this aspect of life are generating highly productive spaces within which individuals are encouraged to engage in creative consumption. In China, then, it might be argued that fashion is at the far end of a spectrum between state and consumer agency in creative and cultural production.

COPYRIGHT AND THE BUSINESS OF CULTURE IN A DIGITAL AGE

One of the things that makes the story of China's post-reform film, music and fashion industries so complex, and so interesting, is the growing tension between expanding international frameworks for the protection of intellectual property rights and new potential for the use and distribution of copyrighted works associated with transformative technological change. The tension between new technologies that facilitate dispersed copying, distribution and creativity and a desire to protect the rights of creators and promote investment and trade in creative products is not unique to China. However, in China a balance between protecting the rights of copyright owners and promoting the interests of society must be achieved in the context of very real development goals, as well as the challenges associated with the transition from a closed, centrally planned economic system towards a much more internationally integrated, market-driven model.

In the second half of the twentieth century 'core copyright industries' such as film and music were associated with a rhetoric that asserts that high levels of copyright protection are crucial to the existence of an economic contribution made by creative and cultural sectors of the economy (Boyle 2004). However, countervailing views about the benefits of intellectual property protection have long existed within the literatures of both law and economics.[1] While international trades in cultural and intangible goods now form an increasingly valuable portion of the global economy, the concepts upon which copyright relies – 'creativity' and 'originality' – are difficult to transplant into new cultural, economic, artistic, social and technological contexts (Alford 1995; Mertha 2005; Liu 2006).

Significant costs are associated with the expansion of global intellectual property frameworks, particularly for developing countries that export few copyrighted products but require access to a great many in order to meet their development goals. At the same time, new technologies are helping to break down physical barriers previously associated with creation, distribution and use. In so doing, these technologies often encourage the use of content in ways that violate the rights of copyright owners. Given the costs associated with overly restrictive copyright protection for developing countries and the challenges for copyright associated with digital transformations, it is important to critically examine the extent to which copyright protection promotes creative innovation and growth in the creative industries.

In 2001 China signed the Agreement on Trade-Related Aspects of Intellectual Property Rights (TRIPs) – a requirement of World Trade Organization membership. The associated amendments to its copyright law have done a great deal to bring China into line with Organisation for Economic Co-operation and Development (OECD) copyright norms. In spite of this, levels of enforcement remain extremely low, and estimated rates of 'piracy' in film and music are as high as 90 per cent (Kennedy 2006). Although few figures are available on rates of trademark violation within China, it is widely accepted that the sale of 'fake' branded fashion items is also very common. China faces heavy pressure from foreign copyright owners and governments eager to expand markets for creative and cultural goods to better enforce its intellectual property laws. Nonetheless, creative and cultural businesses operating in China often have little choice but to adopt strategies that are not overly reliant on intellectual property enforcement.

The PRC's first copyright law did not come into existence until 1990. This was the beginning of a decade in which digital technologies, personal computers, mobile communication and the internet would transform cultural and communicative landscapes globally. The spread of new technologies for the copying, communication and use of content and very low levels of copyright enforcement have made it difficult for creative industries' business models that rely on an ability to control unauthorized copying and distribution of physical media to take hold in China. As a result many creative and cultural entrepreneurs have been prompted to explore new approaches to the value of creative products, and the ways in which their production might be financed and commercial returns generated.

This book draws heavily on in-depth interviews with filmmakers, musicians and fashion designers, as well as entrepreneurs, lawyers, intellectual property court judges, and cultural officials conducted between 2003 and 2009. In the course of these interviews it became clear that many of China's film, music and fashion businesses are actively experimenting with strategies that allow them to function in a weak copyright environment and to operate commercially within limitations on content and industry structure set by the state. Both businesses and consumers are adapting to the regulatory environment in which they find themselves and finding ways to navigate highly complex landscapes of law, policy and commercial opportunity to maximum advantage.

While a large 'grey economy' exists around the distribution of unauthorized content, there is also growing awareness of the scalable value associated with intellectual property rights among Chinese creative industries professionals. At present, many of the businesses that are making money most successfully in the new environment are doing so by providing services, rather than as content owners. There is growing tension between groups that are developing new approaches based on services and advertising (particularly search engines) and businesses such as record labels, which continue to depend heavily on an ability to trade copyright. Copyright owners are beginning to pursue enforcement through the courts, and 'rent seeking' is by no means universally accepted as a legitimate way to make money.

According to Ronald Bettig (1996), copyright has the potential to act as a mechanism for converting creative works into instruments for the expression of capital 'much like real estate, bonds, stock, licenses, franchises, precious metals and so on' (Bettig 1996: 36). Political economists emphasize the expansionary logic of capital as a driving force in the formulation of intellectual property policies. As Bettig puts it:

> when it comes to the domains of information and culture, the logic of capital drives an unending appropriation of whatever tangible forms of intellectual property and artistic creativity people may come up with, as long as this creativity can be embodied in tangible forms, claimed as intellectual property and brought to the marketplace. (Bettig 1996: 34)

Thus, it is to be expected that groups whose property entitlements and power over the market are increased by the expansion of intellectual property rights are strongly supportive of the copyright system. In markets where the copyright system is well established, record labels, publishers and vertically integrated film corporations frequently act as mediators between creative labourers and the market, coordinating the supply of new content and creative products with markets, providing the capital resources necessary for the physical production of creative works and making investments in promotion in order to stimulate demand. These coordinating firms depend heavily on the existence and enforceability of assignable intellectual property rights in order to define the products in which they invest and trade.

Nations in which the production and trade of copyrighted material make a significant contribution to the economy, such as the United States and the United Kingdom, have a similar vested interest in increasing international markets for their products through the construction of international frameworks for intellectual property protection capable of facilitating the international trade of copyrighted material. As commercially focused creative and cultural industries within China develop, commercial classes with an interest in protecting and expanding their entitlements are also forming. The growth of commercial classes with an interest in higher levels of copyright protection and enforcement may well serve as a vital element in the emergence of a stronger copyright regime in China. China's copyright authorities have recognized the significance of this group in the development of the nation's copyright system and are actively exploring ways to stimulate the growth of a copyright market. In 2009 the Copyright Protection Centre of China launched the world's first national copyright exchange: a live platform for trade in copyrighted works (*China Daily* 2009). Although the platform is still in its infancy, it is a deliberate effort to stimulate awareness of the commercial value of copyrights and to encourage investment in the creation of new works.

But larger questions remain about the extent to which the model of copyright protection posited in international frameworks such as TRIPs will benefit the growth of knowledge, the interests of society and, in the twenty-first century, the emergence of industries that take advantage of the communicative and collaborative potential of new technologies. As Bettig also observes, an ability to control the means of communication is central to the way in which 'capitalist classes' use copyright to extract value from artistic and intellectual labour in market-driven economies (Bettig 1996: 35). In China the state continues to maintain a high level of control over important distribution mechanisms, including print and broadcast media, as well as cinemas and mobile communication networks, and its efforts to control content available through the internet have been well documented. As a result, commercial classes have strong incentives for working within (and often around) the boundaries defined by the state.

The challenge for China's policymakers of harmonizing laws and regulations, establishing legal and administrative frameworks, negotiating the role of independent associations in representing the interests of specific groups, and cultivating a culture among entrepreneurs and regulators that supports intellectual property rights is formidable. It is unsurprising that it is taking time for legal frameworks intended to support highly specific models of market activity to supplant older systems of cultural sector management. Although China has formally adopted a copyright law that accords, broadly, with international standards for protection, it remains interested in controlling access to news and information by the public, and concerned with the impact of heterodox content on the social and political stability of the nation. New technologies and new ways of creating, using, distributing and monetizing content and culture are adding to the complexity of both limiting access to unauthorized content and ensuring that intellectual property laws are enforced. Regulators are struggling to keep pace with innovations being made by businesses, cultural producers and consumers.

In the case of film and music in China, both the business sector and the state have reasons for encouraging consumers back into centrally coordinated systems of distribution in China. However, users in China as in other markets are also motivated to actively seek out the content that they enjoy at the lowest possible price. Many creators who do not have access to official distribution channels are eager to share their work with others – either for fun, as part of a status or esteem system, or in the hope of financial gain. Increased access to physical technologies of copying, distribution and consumption of film and music and the growth of both peer-to-peer and grey-market distribution networks are challenging the state's ability to directly control what its citizens are able to access. They have also made it difficult for emerging commercial and entrepreneurial classes to use copyright to control the terms on which works might be distributed and used, prompting criticism of China's weak copyright environment, as well as experimentation with business models capable of functioning in the absence of controlled distribution.

GOVERNMENTALITY

China's policymakers remain keenly interested in popular culture and are developing methods of exerting their influence in the context of a market-driven economy. The internet, publishing, print and broadcast media are all subject to policies aimed at preventing content deemed damaging to the interests of society or the state from being distributed.

Efforts to influence the moral, social and economic frameworks in which people function have taken on a prominent role in strategies for governing creative and cultural areas of production and consumption in the context of economic reform. Licences are required for film and television production, and official approval is needed if a film is to be screened in a legitimate cinema. Concerts and public performance are also controlled. Although China's policymakers are not able to remove unauthorized content from the media-sphere completely, they are able to restrict access to opportunities for the commercialization of any content that has not been approved by the authorities.

In this context, Michel Foucault's insights into the processes involved in modifying the behaviour of a population help cast light on what the state is attempting to accomplish and why current developments are interesting. As Foucault observes, 'disciplining societies' involves more than the exercise of corporal power over individuals: locking people up if they break the law or placing them in a position where they have no choice but to comply with the will of the governing, for example removing all possibility of watching unauthorized content by ensuring that no unauthorized content is physically available. Rather, individuals' ability to choose whether to comply with the will of those who govern them is an essential element of any relationship of power. Foucault describes power as something that:

> operates on the field of possibilities in which the behaviour of the active subjects is able to inscribe itself. It is a set of actions on possible actions; it incites, it induces, it seduces, it makes it easier or more difficult; it releases or contrives, makes more probable or less; in the extreme, it constrains or forbids absolutely, but it is always a way of acting upon one or more acting subjects by virtue of their acting or being capable of action. (Foucault 2002: 341)

According to Foucault:

> What is to be understood by the disciplining of societies in Europe since the eighteenth century is not, of course, that the individuals who are part of them become more and more obedient, nor that all societies become like barracks, schools or prisons; rather, it is that an increasingly controlled, more rational, and economic process of adjustment has been sought between productive activities, communications networks, and the play of power relations. (Foucault 2002: 341)

This statement accurately describes what is now taking place in China, as efforts are made to encourage both consumers and producers to operate within the moral and political boundaries set by the state. Economic reforms, combined with technological developments, have made it harder to control access to unauthorized content. At the same time creative citizens and commercially focused cultural production are now seen as desirable. Private investment and innovation are required to ensure that controlled spaces are able to compete with unauthorized content for the attention of consumers. Encouraging entrepreneurial behaviour and commercially oriented cultural production within carefully delineated moral and political boundaries is thus playing an important role in disciplining China's post-reform creative industries.

In this book, I suggest that Foucault's concept of governmentality, which describes the role that individuals play in their own regulation and control through the internalization of standards of behaviour, is useful in understanding changes in the ways the state, individuals and commercial actors relate to and interact with creative and communicative works in China in the twenty-first century. Governmentality allows us to consider how the relations that typified the socialist control state are being modified in the light of incremental reforms, moving cautiously towards the model that has been called 'authoritarian liberalism' (Jayasuriya 2001; Keane and Hemelryk Donald 2002).

Rather than focusing on the use of techniques of power and domination to control individuals, governmentality draws attention to the relationship between techniques of government and techniques of the self in shaping, guiding, modifying and correcting the ways in which individuals conduct themselves (Burchell 1996). Understood through this framework, intellectual property law, particularly copyright and trademark, represents an important point of contact between mechanisms of domination – the coercive power of formal law, and techniques of the self: self-regulation by both individuals and industries that rely on access to creative works, as well as the legally granted rights associated with creative works and identities.

Inevitably, intellectual property law also represents a point of intersection between the Chinese government's desire to exercise the power of the state within its own borders and its need to comply with the international frameworks of a global trading system. Dynamics of power between the state, firms and citizens include pressure on the Chinese government from foreign governments, as well as lobby groups acting on behalf of international copyright- and trademark-driven industries and commercial interests within China itself. It is well documented that US-based film and music industry lobbies have influenced the current international protection framework (Arup 2000; Wang 2003; Miller *et al.* 2005). Few commentators contest the proposition that the copyright law now in place in China reflects values and expectations established in the United States and Europe. While China's copyright legislation now accords with OECD norms, debates about the most appropriate levels of protection within any country continue. Scholars and activists from around the world have maintained their call for reform of the copyright system (Royal Society for the Arts 2005).

THE ROLE OF LEGISLATION

Since Mao's death, China's leaders have placed an increasing emphasis on the importance of formal law as a tool of government (Peerenboom 2002). However, as Seidman and Seidman (1994) observe, 'inevitably, people choose how to behave, not only in response to the law, but also to social, economic, political, physical and subjective factors arising in their own countries from custom, geography, history, technology and other, non-legal circumstances' (Seidman and Seidman 1994: 45).

Legislative frameworks are just one factor in the complex landscape that consumers and businesses must navigate as they choose how to invest money, effort and time. It is thus inevitable that local circumstances, including distribution networks and social and political environments, will have a profound impact on perceptions and implementation of something as abstract as intellectual property rights. Social factors play an especially important role in markets for creative and cultural products, where 'value' and 'quality' are both subjective and contextual.

Capitalizing on social aspects of consumption by providing brand identities that can be co-opted by consumers is a well-established marketing strategy. A similar strategy can also be used to create negative associations around certain behaviours, for example campaigns intended to discourage consumers from purchasing 'pirated' films or music, or wearing 'fake' designer labels. Moral rhetoric is perhaps one of the most powerful tools in this kind of strategy. The Chinese government has been funding campaigns intended to establish a relationship between piracy and morality for a number of years (Wang 2003). In Beijing, for instance, it is now common to see television advertisements and billboards sporting slogans such as 'Be a responsible Beijinger; don't buy pirated films.' These campaigns are helping to build an image of copyright infringement as 'immoral': an important step in establishing it as a proscribed activity and employing 'techniques of the self' (Foucault 1988: 18–19) in controlling behaviour.

Employing moral rhetoric to mark out behaviour as not only illegal, but also immoral, is a well-recognized technique for maximizing the law's impact. As discussed earlier, the state's power over individuals involves much more than simply creating laws and punishing those who violate them. Imparting the values of the governing to the governed is a defining feature of 'hegemony': 'hegemony is a state within society whereby those who are dominated by others take on board the values and ideologies of those in power and accept them as their own' (Mills 2003: 75).

As Foucault points out, the successful government of others also depends on the capacity of the governed to govern themselves. If governments are to avoid the extremes of domination, they must aim to recruit the conduct of the governed – that is, to operate through a population's ability to regulate its own behaviour. Barry Hindess summarizes Foucault's position: 'successful government of others is often thought to depend on the ability of those others to govern themselves, and it must therefore aim to secure the conditions under which they are enabled to do so' (Hindess 1996: 105).

Encouraging the governed to govern themselves and ensuring that acting within the law is consistent with maximizing individual utility have important implications for the ways in which cultural production is managed in China. Ensuring that filmmakers and musicians accept the limits within which creativity and innovation are considered desirable by the state and convincing consumers that such limits are reasonable and compatible with satisfying cultural and entertainment experiences are more important than ever before in an era of dispersed productive capacities and distribution networks. The Chinese government's ability to enforce intellectual property law also raises important questions about its ability to control its own ranks – to limit corruption, to bring local officials into line and to ensure that centrally promulgated laws and policies are put into effect as is intended by the central government.

ENTREPRENEURIAL GOVERNMENTALITY AND COPYRIGHT

But, as I have already mentioned, there is more to this story than the exercise of state power over individuals. Efforts to integrate China into the international trading system, pressure by international actors on the Chinese government to build and implement a framework for the protection of intellectual property in the creative and cultural industries, and the Chinese government's own efforts to influence the behaviour of citizens are all important. However, the recognition of intellectual property rights associated with creative work and the growth of a highly competitive consumer landscape are prompting deeper shifts in the way that some groups in China think about creative works and activities. A market-driven economic system is creating opportunities for individuals to begin thinking about and acting upon creative and cultural products in an entrepreneurial manner.

Alexei Yurchak's extension of Foucault's concept of governmentality in relation to the activities of entrepreneurs is particularly helpful in understanding this process. According to Yurchak, writing about transitional economic activity in post-socialist Russia, to be an entrepreneur is to have an 'entrepreneurial governmentality', a disposition that allows entrepreneurs to understand economic and social relationships in terms of symbolic commodities such as risks, capital, profits, costs, needs and demands (Yurchak 2002). Russia and China are both making the transition from a centrally planned economy to one that is market driven. Many of the points raised by Yurchak about the challenges of building an entrepreneurial culture in a system that has not formally encouraged entrepreneurialism relate equally to China.

Yurchak explores the relationships between entrepreneurial activity and corruption in an environment where state-sanctioned channels have failed to meet the needs of consumers. Yurchak's concept of entrepreneurial governmentality helps to explain the changing ways in which creative producers and investors understand what they are making and how it might be used in China. In the case of film and music, the formal introduction of copyright law and market-driven reforms of the cultural sector are prompting individuals and groups involved in the production and distribution of film and music products to begin thinking about creative works as assets, which can be invested in, sold on and rented out, and which have the potential to generate profits.

An ability to understand economic and social relationships in terms of symbolic commodities such as risks, capital, profits, costs, needs and demands is not exclusive to creative industries professionals. As the fashion industry demonstrates, entrepreneurial governmentality is also the domain of consumers. Since Mao's death, Chinese consumers have become avid participants in the global fashion system: an area of consumption in which a consumer's investment of time, money and creative experimentation is part of a game that involves both risk and reward.

FROM COPYRIGHT INDUSTRIES TO SOCIAL NETWORK MARKETS

In markets such as the United States and Western Europe the film and recorded music industries developed during an analogue era, alongside an expanding system of intellectual property rights. In these markets copyright law often acts as a mechanism linking copyright owners with distributors and consumers and helping to coordinate a relationship between popularity and income for creators and investors. Commercial film and music businesses in China, on the other hand, are grappling with both a weak copyright system and the impact of digital technologies on their ability to control the copying and distribution of creative works. Ensuring that only authorized copies of a film or music product are distributed and that the copyright owner receives correct payment for each copy sold is all but impossible. In Bettig's terms, copyright owners are not able to control the communication infrastructure in China (Bettig 1996: 2).

Technology, rising disposable incomes and a liberal-market economy are combining to create new possibilities of thought and action for ordinary people, allowing them to adopt a much more active role in both the production and the consumption of creative and cultural works. The values that allow the fashion industry to function – a 'risk' culture, honorific values associated with the *use* of creative and cultural products rather than more passive processes of *consumption*, and the incorporation of creative and cultural products into processes of identity formation, self-expression, communication and play – are quickly becoming evident in other areas of the creative and cultural industries.

In order to survive, film and music enterprises are being forced to understand what they have to sell in new terms, and to seek out new ways of making money, and new business models and strategies for expanding the regulated spaces in which they operate. The relationship between the state, commercial classes and consumers is being tested and negotiated. Creative and cultural industries must now satisfy the demands of the market in addition to those of the state. There is tension between businesses' desire for regulated space and the government's limited capacity (or will) to provide this.

According to Ruth Towse:

> Creativity is central to the cultural or creative industries. It plays the equivalent role in these industries to that of innovation in other sectors of the economy. Just as firms in manufacturing have outlays on research and development (R&D), so firms in the cultural industries search for new ideas and talented workers – artists – to create and supply them. (Towse 2001: 1–2)

New possibilities for interaction with creative and cultural products, digital technologies and instant communication are allowing users to become active participants in processes of production, distribution, creative experimentation and the selection of talent that were previously the domain of firms and commercially driven entrepreneurs. Amateur users and creators are being prompted to invest time and thought in choosing what and how to consume and actively seeking out skills, information and creative resources that allow them to derive maximum benefit from their consumption choices. It is also becoming possible for creativity to be sourced and coordinated among whole populations, rather than depending on more centralized processes of creation and distribution.

Invidious distinction (Veblen 1899), kudos, self-expression, social interaction and a desire for 'fun' are important drivers of investments of time, money, creative labour and risk among increasingly active 'consumers' in China and the rest of the world. Although established systems of production and market coordination still play an important role in China's emerging creative industries, entrepreneurial consumers are becoming central players in the generation of new creative knowledge and contributing to change and growth in the economy in important ways. As Potts (2003) observes, entrepreneurialism, knowledge and the democracy of economic agents are three driving forces of economic growth:

> Competitive or entrepreneurial actions create new knowledge and/or destroy old knowledge, and the market – the democracy of economic agents – decides whether or not it is a good idea. People are motivated by private gain, but if they succeed, then it becomes a public gain: an old problem is better solved, or a new problem is solved. This is what entrepreneurs do, and it is why they are central to the health of an economic society. Entrepreneurs drive economic evolution, and thereby, if harnessed, economic growth. (Potts 2003: 4)

An increasingly open and dynamic commercial and cultural environment in China is giving rise to new opportunities to explore and express identity and values through clothing, entertainment, hobbies, food and experiences, particularly in urban settings. Many Chinese consumers are now eagerly seeking out the literacies and skills necessary to navigate complex media, commercial, social and cultural landscapes. The desire for information about how to live, how to consume, what brands and symbols represent and which purchasing choices or investments of time and effort will provide the greatest rewards in the specific context that an individual inhabits is creating new markets for information.

In the face of so much choice, social network markets (Potts *et al.* 2008) are coming to the fore. Social network markets allow individuals to take advantage of the tastes and intelligence of friends and associates and, with the help of information networks such as the media, to glean information from the decisions of consumers in other cities or countries. Recommendations from trusted others allow individuals to minimize the risks of their own purchases and can serve to integrate an individual more firmly within a peer group. Social networks are also more dynamic than many other sources of information, and capable of providing up-to-date knowledge that accurately reflects access options available to users, regardless of whether they are legal or illegal, or part of the free economy, the grey economy or a formalized economic system.

It is arguable that, in actively seeking out creative products and information from the widest possible range of sources, both legal and illegal, and exploring new ways to interact with creative content, Chinese consumers are acting as catalysts in processes of creative evolution. Their participation in markets for creative and cultural products is a form of transformative engagement, which has an impact on both the form of creative products and the organization of the productive processes and business models that supply them. In Beinhocker's (2006) terms, their activities both feed and shape the evolutionary dynamics of knowledge growth and wealth creation. Active consumers are a powerful source of creative energy and population-wide experimentation involved in product innovation, and are shaping the markets within which physical technologies, business models and regulatory frameworks must function.

While governments and sectors of the established copyright industries attempt to regulate consumer behaviour through the use of moral rhetoric and formal law, individuals are actively developing and applying their own knowledge of technical possibility and practical boundaries to their daily lives. Users are exploiting available resources to take control of distribution processes, ignoring formal legal restrictions relating to reuse of content and developing and applying their own creative and cultural codes to their actions. More and more, the lived reality for many Chinese citizens involves cultural and creative curiosity, expressed as a desire for new experiences and new products, as well as a desire to play and create.

The internet and other communication and copying technologies are disruptive technologies precisely because they have superseded networks of production and distribution previously dominated by either governments or established corporate elites. The difficulties associated with applying business models based around linear value chains of professional production of creative works, controlled distribution and limited possibilities for use in China are allowing new modes of creative production and interaction to evolve. Business models in China are also evolving as firms adapt to the environment in which they must operate and compete to win the attention of ever more worldly, demanding, fashion-conscious consumers.

Copyright law represents a point of intersection between the coercive power of formal law and self-regulation by individuals and copyright industries. However, the spaces that now exist between formal regulation and possibilities of action are, arguably, producing the highest levels of innovation in both the uses of creative works and the generation of new business models. Low levels of copyright enforcement, access to new technologies and an enviable range of content are allowing both consumers and creative entrepreneurs to begin thinking about creative works in new ways. Content is being understood as something that can be chosen, enjoyed on a range of platforms, manipulated, compiled, given to friends and used as a fashion item or social accessory. Rather than something that professionals make and distribute for non-professionals to consume, content is becoming something that can be made, distributed and remixed by anyone. Both creative and cultural industry professionals and users are beginning to understand the value of content in terms of its social functions, the services and accessories that allow it to be delivered with the highest levels of convenience or novelty and its potential as a resource for reuse, remixing and re-creation.

The gaps between formal law and practical reality are prompting businesses to explore new approaches to capturing the commercial value of creative works. However, they are also allowing consumers to explore new ways of using and interacting. It is conceivable that, rather than making the transition towards better enforcement of copyright law and the emergence of copyright-driven creative and cultural industries, new approaches to the role of value in the creative industries may be emerging in China. China's emerging commercial classes are beginning to understand creativity and content in terms of their value as intellectual property. But they are also becoming aware of the attention economy, the power of social network markets and the possibility of charging for services, such as advertising or the delivery of content to mobile devices, rather than for the content itself. The future of business models based on the ownership and controlled distribution of expert-generated copyrighted material is far from assured.

CHAPTER OUTLINE

As mentioned earlier, in order to better understand this process of evolution away from government by the state in China, towards creative and cultural industries in which entrepreneurial consumers are driving forces, this book explores the development of three very different creative industries: film, music and fashion. These three industries are examined through the lens of changing approaches to governance in China's cultural sector, as well as the role that intellectual property is playing in both this transformation and the growth of China's creative industries. The film, music and fashion industries all rely heavily on creative professionals who apply their skills and talents to developing products that are sold in highly competitive markets. However, as this book explores, there are also important differences between the ways in which the power of the state is being exercised in relation to these industries in China, and in the role of intellectual property in their growth and development.

Chapter 2 explores China's integration into a global system of intellectual property rights and critically examines arguments that high levels of 'piracy' in China are a result of a clash between foreign approaches to protecting intellectual property and Chinese attitudes to creativity and copying. The chapter also further explains how the concepts of discourse, governmentality and entrepreneurial governmentality might help to shed light on debates about intellectual property protection in China and the Chinese government's approach to its implementation. Finally, Chapter 2 highlights the difficulties inherent in widely accepted definitions of the creative industries, which emphasize intellectual property, and attempts to understand and promote growth of this sector of the economy in developing nations.

Chapter 3 focuses in on China's film industry. A growing proportion of China's population now has access to cameras, personal computers and broadband internet connections. Technology is making it possible for more people than ever before to engage in what Xu (2007) calls 'cinematic modes of production'. However, the commercially oriented film industry continues to depend heavily on access to coordinated mass distribution channels, as well as the relatively high levels of capital, specialist skills and equipment required for the production of feature films. The commercial film industry's dependence on access to high levels of investment and a coordinated cinema distribution system have made it possible for the Chinese government to maintain a high level of control over this area of production, in spite of the fact that new technologies have made controlling the public's ability to access content difficult.

Chapter 4 considers the situation in the music industry, where the state's power is being exercised less directly than in the case of film. Although some groups estimate that up to 90 per cent of physical music products are 'pirated' (Kennedy 2006) and copyright enforcement online is a major challenge, rapid mobile uptake has been associated with the growth of a lucrative market in mobile music, which is quickly becoming China's most significant source of music-related revenue (Yao 2007). But, while commercial interests, private investment and consumer taste are playing a vital role in the formation of an entirely new market for music products sold directly to mobile devices, China's government remains a dominant stakeholder in the nation's mobile communication networks.

In Chapter 5 it becomes clear that the state is playing an entirely different role in the growth of China's fashion industry. Since the very earliest days of reform and opening up, fashion has been associated with a level of creative freedom absent in many other areas of cultural production in China. The fashion system's dependence on social networks and consumer engagement in order to function has made fashion particularly resistant to attempts to directly influence what people wear. The government's focus in the post-reform era appears to have been on creating an environment conducive to the growth of a market for the products of an industrialized fashion system, rather than on attempting to directly control what is designed and worn. This approach, which has coincided with market-oriented reforms of magazines and the growth of an advertising industry, seems to be working particularly well in China. Consumer demand for genuine versions of branded products appears to be countering the availability of 'fake' products, and China's fashion industry is developing quickly.

Chapter 6 considers the troubled relationship between economic approaches to the creative industries and intellectual property theory. Written with evolutionary economist Jason Potts, this chapter argues that much of the confusion about intellectual property's role in the creative industries results from overlooking three important aspects of the relationship between IP and the growth of knowledge. These are: 1) the effect of globalization; 2) the value of monopoly incentives to create input, compared to the value of reusing creative output; and 3) the evolution of business models in response to institutional change. We argue that the evolutionary dynamics of these three factors suggest that a substantial weakening of intellectual property will, in theory, produce positive net public and private benefit.

Finally, Chapter 7 considers important areas of research yet to be explored in relation to both the growth of the creative industries in China and the role of intellectual property law in the creative industries more widely. Given the significance of the creative industries to the strength of the global economy, and their potential to help developing nations shift away from their dependence on low-cost manufacturing toward higher-value-added areas of creative production, understanding the mechanisms for their growth may provide important clues about the regulatory frameworks and policy approaches likely to help them succeed.

Overall, the book shows that social network markets and consumer creativity and entrepreneurship are powerful forces in the production and commercialization of cultural commodities in the twenty-first century. As the case studies explored in the book demonstrate, entrepreneurial consumers with access to content from a diverse range of sources are helping to drive both creative and commercial developments in China. Although copyright is playing a role in the creative economy that is emerging, China's experience demonstrates that intellectual property protection is not a prerequisite for the growth of the creative industries. Rather, it is just one factor in the dynamic co-evolution of physical technologies, social technologies and business models. Thus new approaches to understanding the economic relationship between intellectual property protection and the creative industries, which recognize the dynamic nature of business models and the productive and entrepreneurial value of creative consumers, are needed.

NOTE

1. An excellent discussion of this history can be found in Deazley (2006).

2. Dynamics of power: from state to consumer

Since the establishment of the World Intellectual Property Organization (WIPO) in 1970, the breadth, scope and terms of global intellectual property protection have expanded steadily (Boyle 2004). As Miller and others discuss in *Global Hollywood 2* (2005) a significant proportion of the world's media and cultural products are now made and consumed transnationally. Governments around the world are searching for ways to encourage innovation and to add value to their economies by fostering newly defined 'creative industries'. At the same time, digital technologies and new possibilities for peer-to-peer distribution and consumer co-creation are blurring boundaries between 'producers' and 'consumers' (Bruns 2008; Hartley 2009) and challenging established business models in what have traditionally been understood as 'core copyright industries'. As policymakers, businesses and users jostle to optimize their interests in a rapidly changing economic, technological and global landscape, debates about how the rights of those who invest in creative products should be protected, and how the benefits to society of creativity and innovation might be maximized, have taken on new significance.

The first section of this chapter introduces some of the most important debates surrounding China's integration into rapidly developing global intellectual property frameworks, focusing in particular on the two areas of intellectual property most frequently connected to the creative industries: copyright and trademark. China has been the subject of criticism from trading partners for its failure to comply with foreign standards of protection for intellectual property rights (IPRs) for more than a hundred years (Alford 1995; Grinvald 2008). A great deal of progress has been made in developing a legal framework for the protection of intellectual property rights since initiatives to reform China's economy and increase international trade began during the late 1970s. However, low levels of enforcement remain a significant source of tension between China and established exporters of intellectual property rights, and historical factors continue to influence the ways in which this class of right is understood and implemented within China.

Furthermore, as the section 'IP and developing countries' discusses, some commentators have questioned the benefits to developing nations, including China, of compliance with a system which many believe favours the interests of established exporters of intellectual-property-related products (Chang 2008). It is widely accepted that China's goals in developing an intellectual property system are closely linked to its desire to gain access to the wider benefits of participating in an international community of trade and encouraging foreign investment and technology transfer. Even within nations in which the role of intellectual property rights in the production and distribution of creative and cultural content is well established, a growing body of commentators suggest that overly restrictive IP systems favour key interest groups at the expense of public access to knowledge and content, and are preventing the creative and collaborative potential of new technologies, such as the internet, from being fully realized (Lessig 2001, 2004).

Regardless of such critiques of intellectual property's expansion, commercially driven reforms of cultural production systems and the introduction of a new class of property rights related directly to content and culture are prompting important changes in the ways in which both producers and consumers are operating in China. It is here that Alexei Yurchak's (2002) extension of Foucault's concept of governmentality becomes useful – providing a framework through which the emergence of entrepreneurial modes of thinking among both producers and consumers of cultural content in China might be understood. Finally, this chapter briefly considers the ways in which an emphasis of social network markets and the entrepreneurial tendencies of consumers might influence an understanding of the role of intellectual property within China's emerging creative industries.

IP'S INTRODUCTION TO CHINA

Global trade in content and culture is big business. Cultural and creative industries have been estimated to account for 7 per cent of the world's gross domestic product (GDP) (UNESCO 2005: 9). Trade in cultural goods almost doubled between 1994 and 2002: from US$39.3 billion in 1994 to US$59.2 in 2002 (UNESCO 2005: 9). Between 1977 and 1996 US 'copyright industries' grew three times as quickly as the overall economy (Miller *et al.* 2005: 10), and in 1996 'cultural industry' sales – film, music, television, software, journals and books – became the US's largest export.

As the value of intellectual-property-related trade has increased, countries that wish to participate in global networks of trade have faced growing pressure to adopt higher standards of IPR protection. It is widely accepted that a few developed nations which are responsible for the bulk of intellectual-property-related exports, most notably the United States and members of the European Union, have driven the global expansion of intellectual property rights. Formally identifying intellectual property as an issue of trade, rather than of morality or social justice, has been central to this process. In many instances, access to the benefits of international trade is now conditional on protection for intellectual property rights. As Maskus (2000) notes:

> In 1984, the United States designated inadequate protection of patents, trademarks, and copyrights as an unfair trade practice that could invoke retaliation under Section 301 of the Trade Act of 1974. In the ensuing 16 years, intellectual property rights (IPRs) have moved from an arcane area of legal analysis and a policy backwater to the forefront of global economic policymaking. Indeed, the world is witnessing the greatest expansion ever in the international scope of intellectual property rights. (Maskus 2000: 1)

In the case of China, economic reform and efforts to promote international trade since Mao's death in 1976 have been associated with steady progress towards the construction of an intellectual property system that satisfies the demands of key trading partners. The Trademark Law of the People's Republic of China was adopted in 1982, followed by the 1984 Patent Law. Finally, in 1990 the PRC's first copyright law came into existence. In 2001, China joined the World Trade Organization – an event eagerly anticipated by international commentators and interpreted by many as a development that would ensure greater access to Chinese markets. As a condition of entry, China became a signatory to the Agreement on Trade-Related Aspects of Intellectual Property Rights (TRIPs), and amended domestic legislation on copyrights, trademarks and patents accordingly.

Considerable resources have been devoted to training lawyers and judges, developing administrative mechanisms for the enforcement of intellectual property rights and establishing dedicated intellectual property courts. But, in spite of the progress that has been made in constructing comprehensive legislative and judicial infrastructures for the protection of intellectual property, levels of enforcement within China remain extremely low. Many estimates suggest that up to 90 per cent of film and music products consumed in China are 'pirated' (Kennedy 2006). Levels of unauthorized copying in areas such as trademarks are harder to estimate, but in 2009 the United States Trade Representative noted that 81 per cent of the infringing goods seized at the US border originated from China in 2008 (Office of the United States Trade Representative 2009: 14). The European Commission reports that 60 per cent of IPR-infringing goods seized by European Community customs authorities originated in China in 2007 (European Commission 2008).

Tension has existed between China and Western trading partners eager to secure protection for foreign intellectual property rights since the early twentieth century (Alford 1995). Prior to the introduction of foreign concepts of intellectual property rights, indigenous systems of cultural production and attitudes to creativity and the legitimacy of copying were well developed. At least two authors, Alford (1995) and Grinvald (2008), argue that attempts to transplant foreign concepts of property rights arising from acts of creativity into China over the past century have enjoyed little success, in large part as a result of failure to take into account local attitudes to copying and Chinese expectations of an intellectual property system.

In *To Steal a Book Is an Elegant Offense*, Alford (1995) advances three broad assertions: that, in spite of the arguments of Chinese scholars and the expectations of Western theorists, China has never developed a sustained counterpart to intellectual property law; that early-twentieth-century attempts to introduce American and European intellectual property law to China were unsuccessful because they failed to consider the relevance of such models for China and instead presumed that foreign pressure would be enough to produce widespread compliance; and, finally, that current attempts to introduce intellectual property law to China have been similarly unsuccessful, as a result of their failure to take into account difficulties of reconciling legal values, institutions and forms generated in the West with the legacy of China's past and the constraints of its current circumstances.

As Alford observes, reverence for the past, often expressed in terms of heavy reliance on copying and learning by heart, was a central feature of China's bureaucratic and educational systems during the imperial era. From the Sui dynasty (581–618 CE) onwards China's rulers relied on a civil service examination to identify individuals worthy of a position in the bureaucracy. This examination system, centred on the Confucian classics, viewed knowledge of the past as evidence that individuals possessed the attributes required to resolve the problems of the present (Alford 1995: 22). Copying functioned as an important source of power and control – the Emperor decided which versions of history were to be learned, which philosophical texts should be memorized and how language should be used.

Alford argues that the heavy reliance on the past inherent in China's systems of morality, education and government was not conducive to a notion that individuals should possess property-like interests in creative works or ideas. On the contrary, the reproduction of particular images or ideas by persons other than the individual who first gave them form provided a means by which the past could be engaged with and applied to the present. Copying from the works of others 'evidenced the user's comprehension of and devotion to the core of civilization itself, while offering individuals the possibility of demonstrating originality within the context of those forms and so distinguishing their present from the past' (Alford 1995: 29).

According to Alford, access to the 'common heritage of all Chinese' and freedom to copy from the canon of Chinese scholarly and artistic works were thus essential to every aspect of China's social and political system (Alford 1995: 19–20).

While copying from the past played a key role within China's philosophical and artistic traditions, Alford suggests that such practices did not preclude artistic innovation or the development of techniques and ideas. Rather than merely attempting to produce facsimiles of the works or decisions of earlier periods, China's artistic, intellectual and legal traditions were centred around a process of 'transformative engagement' with the past (Alford 1995: 25). In Alford's view Confucius exemplified this concept. Confucius was aware of himself as a mere transmitter of knowledge. Nonetheless he was actively involved in the selection and adaptation of this knowledge in meaningful terms to the circumstances of his own time. According to this logic, intellectual endeavour provides the medium through which it is possible to interact with and transmit knowledge long since discerned by one's ancestors (Alford 1995: 22–6).

Similar attitudes to copying and learning from the works of great masters have also existed in Europe and the Unites States at various points in history. It is interesting to note that in their *Introduction to Intellectual Property Law* Phillips and Firth (2001: 127) point out that for many generations European artists regarded the imitation of a classical style or form of expression as the ultimate degree in the attainment of aesthetic excellence. Just as Chinese artists aspired to the form and style of masters who had painted before them, the artists of the Italian Renaissance sought to re-create the artistic achievements of the Greeks and Romans (Luppino 2001). The words of Florentine painter Cennino Cennini (*c.* 1370–*c.* 1440) closely echo Chinese attitudes: 'If you imitate the forms of a single artist through constant practice, your allegiance would have to be crude indeed for you not to get some nourishment from them' (Crofton 1988: 41). Imitation would eventually allow artists to develop a 'good' style of their own: 'because your hand and your mind, being always accustomed to gathering flowers, would ill know how to pluck thorns' (Crofton 1988: 41).

However, copying in China was understood as both more positive and more central to the process of artistic learning and development than was the case in Europe, and reflected a broader philosophical veneration of the past that permeated the highest levels of governance (Alford 1995: 29). In contrast to Europe, where copying of the Renaissance was merely a transient aspect of artistic practice, in China the principles of copying in art were consistently perpetuated and endorsed by state-sanctioned philosophy and actively encouraged by the elite academic nobility, as well as the legal and political systems. Although innovation did occur, it took place against a background of institutionalized acceptance of the central role of the past (Luppino 2001).

In Europe the development of the printing press using movable type, a device that made it possible to realize the economic value of written works, combined with the industrial revolution and increasing rates of literacy, created an environment in which copyright law, and industries which depended on it, could evolve (Phillips and Firth 2001: 127; Marshall 2005: 7–20). Imitation of the styles of past masters gave way to a focus on the 'originality' of artistic creations and notions of romantic authorship (Marshall 2005). The advent of photography also had a profound impact on the context in which art was created. Thus, by 1912 André Salmon is quoted as stating: 'The only possible error in art is imitation; it infringes the law of time, which is the Law' (Crofton 1988: 42).

In the literary world, the intersection between new technology and established industries radically altered the way that creative works were traded and reproduced. In England the publishing industry lobbied heavily for the recognition of proprietary rights in creative works capable of rewarding their investments in new printing techniques. According to Marshall (2005: 7) the printing industry's desire for the recognition of proprietary interests in literary manuscripts did not result from a concern that the rights of authors should be protected. Rather, it was an effort to preserve a monopoly and to ensure that publishers remained profitable in the face of new methods of mass reproduction.

Although various imperial governments in China developed controls over printed works, Alford argues that they were motivated by a desire to control the dissemination of sensitive material rather than to facilitate commerce or promote creativity (Alford 1995: 13). The implication of Alford's work is that China's failure to develop an equivalent to Western-style intellectual property concepts in spite of well-established legal, artistic and commercial institutions represents much more than a mere 'gap' in the evolution of Chinese law. The absence of indigenous concepts equivalent to Western notions of intellectual property rights reflects fundamental differences between perceptions of the role of copying and has important implications for the application of Western notions of intellectual property law in the current era.

Leah Grinvald (2008) echoes Alford's assertion that current difficulties in securing higher levels of protection for intellectual property rights in China are the result of the failure of foreign governments to take into account the historical, political and cultural environment into which new laws are being introduced, as well as the goals and expectations of China's policymakers in creating an intellectual property system. But, while Alford emphasizes the role of Confucian values and artistic traditions in the struggle to apply Western European and American approaches to intellectual property law, particularly copyright, Grinvald focuses on important differences between the emerging theoretical framework underpinning trademark law in China and theoretical approaches to trademarks which have driven legal developments in the United States.

Grinvald argues that the absence of an organic history of trademark protection in China and the circumstances in which recent legislative changes have been made have resulted in important differences between Western approaches to trademarks and those which are driving trademark development in China. Trademark law in the United States draws heavily on the framework of utilitarianism, which Grinvald defines as 'an assessment of the consequences of maximizing the benefits to society as a whole, rather than prioritizing individual benefits' (Grinvald 2008: 63). As Grinvald summarizes:

> Specifically, as applied to trademark law, utilitarianism in American legal jurisprudence justifies legal protection because the protection of trademarks maximizes a benefit to society, namely, reduced search costs associated with the purchase of products. Such protection provides brand owners with incentives to improve the quality of their trademarked products. (Grinvald 2008: 63–4)

Trademarks were the first form of intellectual property to be protected by the PRC, which issued the Provisional Regulations Governing Trademark Registration in 1950, replaced by Regulations Governing the Control of Trademarks in 1963. After 1963 foreigners were also permitted to register their marks in China (Endeshaw, 1999: 33). However, during the Cultural Revolution, the use of trademarks, even as source identifiers, was suspended (Grinvald 2008: 73). Grinvald theorizes that, in contrast to trademark law in the United States, trademark law in China reflects a type of 'social planning theory' with welfare and distributive components (Grinvald 2008: 27). In Grinvald's view, the goals of China's policymakers in reinstating the pre-Cultural Revolution trademark law during the early 1980s and in amending the law to bring it into line with international standards continue to influence the way in which trademarks' purpose is understood by policymakers and the legal community in China.

According to Grinvald, the decision to protect intellectual property by Deng Xiaoping's government resulted from a desire to attract foreign investment and, in so doing, to facilitate economic growth and technological development. An understanding of trademarks as a type of individual property that merits a right to exclusivity and protection from misappropriation lies at the heart of the US approach to trademarks. Not only did China lack an indigenous counterpart to the system of trademark protection that had developed organically in legal systems in the United States and Western Europe, but such an approach was simply not compatible with China's communist political system in the immediate post-Mao period. Grinvald summarizes:

> it was not until recently that it was possible to conceive of trademarks as a type of property right, from either the consumer's point of view or that of the trademark user. As such, the concept of trademarks as a type of asset (as embodied by the goodwill of the trademark) never took root in China. It was not until the reform government of the 1980s that trademarks began to be perceived as a means to improve the economy. (Grinvald 2008: 75)

The absence of strong alternative mechanisms for protecting consumers from sub-standard products in China has also meant that trademarks are viewed as an important method for protecting the health and safety of consumers and, in the case of harm, of fixing liability (Grinvald 2008: 75). Grinvald concludes that a failure by US trade representatives to fully appreciate these differences remains an important factor in the lack of success associated with US efforts to secure higher levels of protection and enforcement for intellectual property.

IP AND DEVELOPING COUNTRIES

Implementing intellectual property legislation demands significant commitments by the state to legal and judicial development, policing and enforcement, as well as to protecting the intellectual property rights of non-Chinese citizens. These are all resource-intensive tasks with serious economic implications for nations such as China which export very few intellectual property products but require access to a great deal in order to meet their development goals.

It is widely acknowledged that developing nations such as China have agreed to the standards of protection and enforcement of intellectual property rights demanded by the TRIPs agreement in order to gain access to the wider benefits associated with World Trade Organization membership and the access to markets which such membership provides (Arup 2000; Maskus 2000). Even the most optimistic assessments of the impact of an expanding global intellectual property framework acknowledge that the greatest burden of change associated with the TRIPs agreement falls on developing nations, whose citizens are required to pay for information and content which would otherwise be freely available. Whether or not protecting intellectual property will enable the growth of domestic innovative and creative capacities remains a key question for China, as it struggles to develop its ability to meet domestic demand for intellectual property products itself, and to move away from its reliance on low-cost manufacturing industries towards higher-value-added areas of production.

Drahos and Braithwaite (2002) are two notable critics of the international norms of information ownership being established through mechanisms such as TRIPs. In *Information Feudalism* they provide a useful discussion of the impact intellectual property developments may have on the production and consumption of knowledge, particularly for developing countries (Drahos and Braithwaite 2002). Intellectual property's role in establishing economic incentives for investment in creativity and innovation forms an important element of theoretical justifications for intellectual property law (Fitzgerald and

Fitzgerald 2004). However, Drahos and Braithwaite (2002) believe that strong intellectual property protection is not necessarily a prerequisite for the development of creative and innovative capacities in developing countries. On the contrary, intellectual property protection raises the costs of creativity and may even retard the development of local IP industries.

Drahos and Braithwaite's concern over the impact of IP's international expansion on prospects for innovation and economic development in developing countries is echoed by a growing number of authors. Ha-Joon Chang (2008) is one such critic. According to Chang:

> the foundation of economic development is the acquisition of more productive knowledge. The stronger the international protection for IPRs is, the more difficult it is for follower countries to acquire new knowledge. This is why, historically, countries did not protect foreigners' intellectual property very well (or at all) when they needed to import knowledge. If knowledge is like water that flows downhill, then today's IPR system is like a dam that turns potentially fertile fields into a technological dustbowl. (Chang 2008: 142)

However, the growing literature on IPRs also includes the work of scholars with more optimistic views about the impact of an expanding global system of protection. Maskus (2000) is one such author, pointing out that massive policy shifts, such as the global expansion of the intellectual property system that has taken place since the early 1990s, inevitably give rise to debate and controversy (Maskus 2000: 6). While Maskus acknowledges that the greatest requirements of change associated with agreements such as TRIPs are placed on developing countries, and that the formation of global systems for regulating IPRs involves real costs:

> such costs can be accompanied by even larger benefits, though with a time lag. Stronger IPRs can usher in more certain contracts that raise the quality of technology acquired and permit tighter partnerships between domestic and foreign firms. They can set the stage for efficient generation of follow-on and adaptive technologies that help diffuse learning throughout the economy. They can provide investments to start up new firms, build product quality, and expand marketing networks. (Maskus 2000: 240)

DISCOURSE, GOVERNMENTALITY AND ENTREPRENEURIAL CONSUMERS

While debates over the benefits of expanding frameworks for the protection of intellectual property rights continue, China's policymakers have committed themselves, on paper at least, to developing and implementing this class of property right. Intellectual property rights are being incorporated into mechanisms of governance and the landscape within which creative production and consumption occur. They are also closely associated with the growing discourse on the creative industries, a policy area that is gaining increasing traction within China as one which might help to shift the nation's economy towards higher-value-added areas of production, and provide a route to capturing growing domestic demand for entertainment, innovation and culture.

French philosopher Michel Foucault's theoretical innovations in relation to the nature of power and the role of discourses in shaping debates and on strategies for governance based on self-examination provide a helpful framework for attempting to unpack the complex relationship between intellectual property rights and the rapidly developing business of culture in China. As Foucault observed, rather than existing as objective truths, notions of justice and morality change over time, reflecting and influencing relationships of power (Foucault [1961] 1989, 1972: 224; Mills 2003: 58). This process is particularly apparent in the debates surrounding intellectual property, the history of which is closely linked to evolving concepts of morality and justice and the impact of a changing economic and technological landscape on conceptions of creativity and processes of innovation.

The language of the debates surrounding intellectual property often reflects the economic and ideological agendas of groups that stand to gain financially from its expansion, as well as those calling for greater attention to equity of access and social justice. On the one hand, groups such as the Motion Picture Association of America (MPAA) regularly refer to 'theft', 'piracy' and 'organized crime', equating stricter enforcement of copyright laws with the growth of the creative economy (MPAA 2004; Treverton *et al.* 2009). On the other, the *Adelphi Charter on Creativity, Innovation and Intellectual Property* refers to the imperatives of human rights and describes the current global intellectual property system as one 'which is radically out of line with modern technological, economic and social trends' (Royal Society for the Arts 2005).

Foucault uses the term 'discourse' to refer to regulated statements that combine with others in predictable ways and which are capable of delineating the boundaries within which debates occur and new possibilities are considered (Foucault 1978: 100–101; Mills 2003:10). Discourses on intellectual property rights are deeply intertwined with the ways that creative individuals, copyright owners, consumers, governments and citizens understand themselves and their relationships with each other and with the world. For example, Alford (1995) argues that Confucian approaches to artistic creation view a creative work as something that has been distilled from a common cultural heritage that belongs to all (Alford 1995: 20). In contrast, the French notion of *droit moral* is based on an understanding of creative works as intrinsically linked to the individual creator and a reflection of the personality of the author (Ginsburg 1992: 15).

Globalization, the advent of new technologies and the rise of multilateral trading systems such as the WTO mean that discourses on creativity and the role of intellectual property rights are now being influenced by transnational processes of production and consumption, as well as consumers who are better networked and more aware of international trends than ever before. It is here that Foucault's concept of governmentality becomes useful. The term 'governmentality' refers to a shift in the operation of power away from the positive exertion of control over the bodies of subjects and towards the active participation of individuals in their own governance. Governmentality[1] describes a process by which norms of behaviour are internalized by subjects who learn to self-censor and self-regulate in accordance with the rules of behaviour generated by institutions such as the family and the school (Mills 2003: 46).

In 'Entrepreneurial Governmentality in Post-Socialist Russia', Alexei Yurchak (2002) extends Foucault's concept of governmentality and applies it directly to the patterns of thought and behaviour associated with entrepreneurialism. According to Yurchak, the term 'entrepreneurship' usually refers to 'the industrious systematic activity of organising and operating a profit-making business venture, and assuming the risks of possible failure' (Yurchak 2002: 278). Gordon summarizes governmentality as 'a way or system of thinking about the nature of the practice of government (who can govern; what governing is; what or who is governed), capable of making some form of that activity thinkable and practicable both to its practitioners and to those upon whom it is practiced' (Gordon, 1991, cited by Yurchak 2002: 279). According to Yurchak:

> In Foucauldian terms, then, to be an entrepreneur is to have *entrepreneurial governmentality* that makes it 'thinkable and practicable' to relate to different aspects of the world – people, relations, institutions, the state, laws – in terms of symbolic commodities, risks, capital, profits, costs, needs and demands, and so on. It is a way of knowing what an entrepreneurial act is, who can act entrepreneurially, and what or who can be acted upon in an entrepreneurial way. (Yurchak 2002: 279)

By recognizing and promoting intellectual property rights in creative products in the context of widespread market-driven reform, China's policymakers are moving towards new strategies for the governance of creative and cultural activities in China. Furthermore, as Chinese consumers become more deeply engaged with increasingly global consumer landscapes, they are being incorporated into global discourses of value, identity and creativity. It is arguable that the creation of an intellectual property system in China since the 1980s has established a framework for the expression of entrepreneurial governmentality in relation to cultural and creative works and activities.

As will be discussed in more detail later in the book, commercially mediated approaches have by no means entirely replaced pre-reform frameworks for the management of cultural production and distribution in China. Nonetheless, the expansion of intellectual property rights has given rise to new categories of items that can be bought, sold, rented and even 'stolen'. Although capitalizing on opportunities associated with legislative recognition of copyright and trademarks in China remains complex, it is clear that creative works have become objects of commerce, investment and proprietary claim, as well as tools of communication, artistic expression and propaganda.

It is not only groups that might traditionally be thought of as 'producers' or 'investors' in cultural and creative products that are acting entrepreneurially in China. There are signs that consumers are also employing entrepreneurial modes of thinking (Hartley and Montgomery 2009) as they seek to navigate an increasingly complex landscape of choice and opportunity and endeavour to maximize the benefits associated with their investments of time, money and creative labour. As this book explores, the extent to which individual creative industries are able to accommodate entrepreneurial consumers varies greatly. Furthermore, the actions of Chinese consumers have implications that are being felt well beyond China's borders. Not only are the global discourses and regulatory frameworks that surround trade in creative works having an impact on laws within China, but China's citizens are being integrated into an increasingly global system of creativity and consumption.

THE RISE OF THE CREATIVE INDUSTRIES DISCOURSE

The rapid international expansion of intellectual property protection over the past 20 years has been accompanied by the emergence of a growing literature on 'creativity', 'culture' and, more recently, 'the creative industries'. 'Creativity' and 'culture' are both complex concepts heavily weighted with notions of identity, nationalism and individuality (Hartley 2005). There has been considerable debate about desirable national policies in these areas, the extent to which it is appropriate to incorporate 'creativity' or 'culture' (or both) into economic policies, and the extent to which it is possible or desirable to transfer terms such as 'creative industries' between cultures (Wang 2004).

The term 'creative industries' softens traditional divides between 'culture', 'the arts' and industry. Rather than separating 'creative activities' and 'the arts' from media industries and the rest of the economy, the creative industries approach is explicit about the role that individual creativity, skill, talent and, most importantly for this book, intellectual property play in driving broader economic growth. The UK's Department of Culture, Media and Sport defines the creative industries as: 'those that are based on individual creativity, skill and talent. They also have the potential to create wealth and jobs through developing and exploiting intellectual property' (DCMS 2010).

The controversial terminological shift from 'arts' and 'culture' to 'creative industries' was first made in the United Kingdom by the incoming Blair government in 1998 (Cunningham 2006: 5). Creative industries policy approaches have since been taken up by governments in East Asia, Australia and New Zealand. Hong Kong has made significant advances in the adoption of a creative industries approach (Centre for Cultural Policy Research 2003; Yung 2003; Hui 2005), and there are signs that mainland policymakers are also being influenced by these trends (UK Trade and Investment 2004; Keane 2007, 2009).

As Michael Keane (2009) observes, the term 'creative industries' first appeared in mainland China in late 2004. Although the language of the 'creative industries' has been taken up with enthusiasm by local- and city-level governments across China and has recently appeared in national-level policy statements, important questions remain about the extent to which its adoption represents a genuine shift away from existing approaches to the role of culture or a commitment to encouraging bottom-up processes of creativity and innovation, particularly among producers of cultural production and content. According to Jing Wang, 'The thorniest question triggered by the paradigm of creative industries is that of "creativity" – the least problematic in the western context. How do we begin to envision a parallel discussion in a country where creative imagination is subjugated to active state surveillance?' (Wang 2004: 13).

Other commentators, such as O'Connor and Xin (2006), have pointed out that the adoption of the terminology of the creative industries may, in fact, represent little more than a convenient rhetorical shift by policymakers eager to portray China as modern and in tune with international developments, and eager to secure the economic benefits of creativity and innovation, but who remain unwilling to relinquish power over key areas of cultural policy (O'Connor and Xin 2006; Keane 2009). Such debates have a clear role to play in understanding the function and usefulness of creative industries policy discourses as they are being integrated into wider cultural agendas and landscapes of reform within China.

However, the experiences of Chinese industries such as film, music and fashion, which are developing in the context of difficult intellectual property environments and a rapidly changing technological landscape, also have important implications for ways in which the creative industries might be understood and defined beyond China's borders. In particular, the capacity of some kinds of business to adapt to an environment in which peer-to-peer communication and high levels of unauthorized copying and distribution are a fact of life may provide important clues about the role of intellectual property in facilitating the growth of 'creative industries', and the value for developing economies of strengthening intellectual property enforcement in an effort to promote the growth of this sector of the economy. It may also provide important insights

into the kinds of business models likely to succeed in the increasingly globalized, interactive landscape of cultural production and consumption of the
twenty-first century.

The 'creative industries' discourse is closely connected with a growing
focus on surveying, measuring and valuing proportions of economic activity
generated by 'cultural' and 'creative' workers. Howkins's (2001, 2005) work
in estimating the size of the 'creative economy' is a clear example of this.
According to Howkins, whose approach broadly coincides with the DCMS
definition of creative industries, the global creative economy was worth
US$2.2 trillion in 2001 and $2.9 trillion in 2005 and would be worth $4.1 tril-
lion in 2010 (Howkins 2005). Howkins defines the 'core creative economy' as
consisting of 15 industries responsible for turning new ideas into new products: advertising, architecture, art, crafts, design, fashion, film, music,
performing arts, publishing, R&D, software, toys and games, TV and radio,
and video games. These industries are an important driver in the global economy, growing at an average of 5 per cent per year (Howkins 2005). As
Howkins observes:

> the growth of the creative economy has meant IP laws, especially copyrights and
> patents, have moved centre stage of the global economy. In the 1980s, IP was a
> marginal factor in most economies and of little concern to most policy-makers.
> Twenty years later it is a central and important factor in almost all economic
> activity. (Howkins 2005: 35)

When the creative industries are defined in terms of creativity inputs and
intellectual property outputs, the necessity of high levels of intellectual property protection and enforcement as a precondition for the growth of this sector
of the economy becomes a truism. Such a definition provides little space for
the evolution of theories that might provide new perspectives on IP's role in
the growth of the creative industries, or approaches to intellectual property that
reflect the major shifts in processes of creativity, distribution and consumption
resulting from new technology. The difficulty of defining the creative industries in terms of creative inputs and intellectual property outputs has been
addressed, to a certain extent, by Li Wuwei, director of the Shanghai Creative
Industries Association and author of *Creative Industries Are Changing China*
[*Chuangyi gaiban zhonguo*], who defines the creative industries as 'those
industries that rely upon creative ideas, skill and advanced technology as core
elements, increase value in production and consumption and create wealth and
provide extensive jobs for the society through a series of activities' (Li Wuwei,
2008, cited in Keane 2009).

Li Wuwei's definition avoids the obvious difficulties for China's policymakers of reconciling a desire to secure the economic benefits of growth in the
creative industries with low levels of intellectual property enforcement and the

developing nature of China's intellectual property system by emphasizing the characteristics of industries that might be classed as creative industries and the benefits to the economy that these industries provide. The issue of *how* these industries turn creative ideas, skill and advanced technology into increased value in production and consumption, create wealth and provide jobs is not addressed, beyond the recognition that this occurs 'through a series of activities'. Such a definition provides few clues about the role that intellectual property might play in facilitating the growth of the creative industries.

CREATIVE INDUSTRIES AS SOCIAL NETWORK MARKETS?

The definition of the creative industries proposed by Potts *et al.* (2008) may provide greater scope for a theoretical exploration of the relationship between intellectual property protection and the growth of commercially focused creative industries. In contrast to the widely adopted DCMS approach to the creative industries as a group of *industries* which involve creative inputs and intellectual property outputs, Potts *et al.* (2008) argue for a market-based definition, which approaches the creative industries in terms of the extent to which supply and demand operate in complex social network markets. A social network market is a market in which an individual's decisions are based on the decisions of others. One example might be a decision to join Facebook, rather than MySpace, because all of your friends are on Facebook, to buy a certain brand of clothing because someone you admire wears that brand or to read a particular book because a friend recommended it.

Potts *et al.* suggest this new definition for several reasons. The first is that the standard industrial classification system, developed over 50 years ago, no longer accurately reflects the much more complex, service-oriented economy of today. The second difficulty of an industrial conception of the creative industries relates to the industrial classification system more generally. That is:

> industries do not actually exist in microeconomic theory. They are not natural categories in themselves. What exists are agents, prices, commodities, firms, transactions, markets, organizations, technologies and institutions. These are what is economically real at the level of an individual agent's transformations or transactions. An industry is a derived concept, and the creative industries doubly so. (Potts *et al.* 2008: 168)

These authors argue that the defining feature of the creative industries is not artistic or creative elements of production, but the role that social network markets play in helping both consumers and producers to navigate highly complex landscapes of taste and value:

The analytic distinctiveness of the CIs [creative industries] rests thus not upon their non-market value, but upon the overarching fact that the environment of both their production and consumption is essentially constituted by complex social networks. The CIs rely, to a greater extent than other socio-economic activity, on word of mouth, taste, cultures and popularity, such that individual choices are dominated by information feedback over social networks rather than innate preferences and price signals. (Potts *et al.* 2008: 169–70)

Such a definition fits very comfortably with the concept of *entrepreneurial consumers* (Hartley and Montgomery 2009) who actively navigate information resources available to them (and who increasingly play a productive role in the creative industries) in an effort to ensure that they are able to choose products and activities that represent the best value for investments of time, money and, in some cases, their own creative labour. For example, in deciding whether or not to see a new film, an individual might find out what others have said about the film on a film-rating website, ask friends or colleagues what they thought of it and consult the opinion of a professional critic in the newspaper.

The decision about whether or not to spend money on a cinema ticket or a collector's edition DVD, rather than downloading the film for nothing, is also likely to be affected by much more than a simple calculation of price. Perhaps going to the cinema might provide an opportunity to impress a girlfriend or to socialize with friends while enjoying a film on a big screen. Owning a legitimate edition of the film on DVD might strengthen an individual's sense of identity as a film fan and provide the focal point for a night in with friends. For some, simply downloading the film for nothing and watching it on the computer and then chatting about it with communities of fellow film connoisseurs, either in person or online, might represent equal 'value'.

The internet has increased the speed with which information about choices can be shared. Music-sharing websites provide opportunities for users to see what their friends are listening to and to try out music enjoyed by members of a social network with the click of a button. Sites such as Amazon are able to show information about what other customers with similar interests have browsed or bought to prospective customers. Social networking sites allow people to share and view vast quantities of information about what they do, where they go and the products and services they consume or desire. But, while the effects of social network markets are being enhanced by new technologies, this form of market has long functioned in many areas of analogue consumption. Fashion, for example, operates as a classic social network market – where the choices of an individual depend on the choices of others. In the case of the fashion industry, analogue media technologies have also developed to provide consumers with information about what others are choosing, guidance about how the rules of the fashion landscape operate, and inspiration to consumers who wish to make innovations of their own.

An important advantage of the social network market definition of the creative industries is its ability to accommodate the rapidly changing, socially driven and difficult-to-predict nature of many creative industries. As Potts *et al.* (2008) point out, these are the very characteristics that distinguish creative industries from other areas of production and consumption, which necessitate constant adaption to changing tastes and demand for novelty and the formation of new business models to adapt to shifting consumer behaviour and new technologies. According to Potts *et al.* (2008) the role of novelty in driving consumption in the creative industries means that consumers rely on social networks for information about quality and value precisely because the product or experience they are considering is new. These areas of consumption are also ones in which producers find it difficult to predict which products will prove popular, and so must offer a range of products in order to manage the risks of investment.

What, then, might an understanding of creative industries as social network markets, rather than as industries in which creative inputs are converted into jobs and economic growth in the form of intellectual property outputs, suggest about approaches to intellectual property most likely to facilitate growth in the creative industries? This is a question that the next chapters of this book will explore.

CONCLUSION

Intellectual property law's development in China has been closely associated with attempts by foreign powers to create an environment 'suitable' for foreign businesses to operate in. As mentioned above, protection of intellectual property rights has been a source of continuing tension between China and its trading partners since the late nineteenth century. The construction of an intellectual property system in the post-Mao period has formed part of much broader processes of economic transformation, changes in the cultural production system and integration into global networks of trade and culture. China's integration into the global economy has also coincided with what Maskus (2000) calls 'the greatest expansion ever in the international scope of intellectual property rights' (Maskus 2000: 1), as well as a period of transformative development in technologies of copying and communication.

Important questions persist about both the benefits to developing economies of protecting intellectual property rights and the role of such protection in facilitating the growth of businesses operating within what are often referred to as the creative industries. Nonetheless, the decision by China's policymakers to recognize and create mechanisms for protecting intellectual property rights and to extend market-driven reforms to areas of cultural production and consumption has created an environment in which both producers and consumers are able to relate to culture in entrepreneurial terms.

A definition of creative industries as social network markets may be particularly useful for developing countries, such as China, which are eager to stimulate and capture the advantages associated with an increase in creative and innovative activity. Although intellectual property legislation has expanded rapidly around the world, and many developing countries have now adopted laws that formally protect intellectual property rights, developing nations face very serious challenges in relation to enforcement. The difficulties associated with enforcing intellectual property law are often accentuated by the spread of new technologies for copying, communication, creation and consumption – such as ever more powerful personal computers, mobile devices and the internet.

In spite of these difficulties, China's policymakers have committed themselves to developing the creative industries, and are seeking out strategies that will allow them to maximize the benefits of investment in the creative economy. Although there can be little argument that business models often associated with highly profitable areas of the creative industries in developed economies depend heavily on strong intellectual property systems, live performance, tourism, fashion, fan cultures and online gaming all depend just as heavily on social network effects. As a result, real possibilities exist for development of creative industries in environments where intellectual property rights cannot be easily controlled.

NOTE

1. Although Foucault did deliver a lecture entitled 'Governmentality' to the College of France in 1978 (Foucault 2002), his exploration of the process of what it means to govern, how power operates and how government of others is achieved takes place across a range of his works. In particular, Volumes II and III of *The History of Sexuality* (Foucault 1985, 1986), 'The Political Technology of Individuals' (in Foucault 2002, pp. 403–17), '"Omnes et Singulatim": Towards a Critique of Political Reason' (in Foucault 2002, pp. 298–326, 'What Is Enlightenment?' (Foucault 1984) and 'The Subject and Power' (in Foucault 2002, pp. 326–48) each consider these themes.

3. China's film industry: tension and transformation

By 2005 China's film industry had been engaged in a process of commercialization and reform since the early 1990s. Policy changes were encouraging private investment in film production (Hui 2006: 63), and filmmakers were required to operate in an increasingly market-driven environment. However, it was clear that, in spite of structural changes to the industry and a shift away from state-funded production, film continued to be viewed by China's policymakers in political and ideological terms. Onerous censorship restrictions, tight foreign film import quotas and a poorly developed commercial cinema network meant that there were major gaps between films that could be legally made and distributed and those demanded by audiences. A weak copyright environment and well-developed illegal distribution networks were two important factors in the popularity of 'pirated' DVDs, which had become the most common form of film consumption for the majority of China's citizens.

For many Chinese filmmakers working at the coalface of the industry's commercial transformation, balancing censorship restrictions with the demands of the market and artistic ambition presented a much more immediate challenge than copyright enforcement. The image that emerges of China's film industry in the first half of the twenty-first century is one of a sector in the midst of transformation – from state-funded production and state-mediated consumption towards much greater reliance on private investment and commercially generated returns, albeit within a context of continued government desire to maintain control over content. This chapter explores this transformation, drawing on interviews with directors and producers working in the mainland film industry in 2004 and 2005, almost 15 years after the enactment of the PRC's first copyright law, and four years after its amendment to meet World Trade Organization (WTO) requirements. Tension between the Communist Party of China's (CPC) view of film's pedagogical and ideological role and established international industry practices, which place a much greater emphasis on the role of copyright as a tool for coordinating the market and controlling distribution, is discussed.

The final section of the chapter touches briefly on the state of the industry in mid-2009. In spite of the difficulties that were confronting the filmmakers interviewed in 2004/05, the Chinese film industry was entering a period of rapid growth. By 2008 it would be reported that revenue had increased by 20 per cent for five consecutive years, and that film in China was 'booming' (Huang Qunfei 2009). In 2009, access to coordinated systems of distribution and revenue collection remained the subject of state policies that emphasized film's pedagogic role and limited access of foreign films to the domestic market. However, the commercialization of film financing and production and reform of cinema distribution systems had created important legitimate spaces for the expression of entrepreneurial activity and powerful economic incentives for making films within the guidelines set by the state.

CINEMA AND THE STATE

On 1 October 1949, Mao Zedong, chairman of the CPC, declared the existence of the People's Republic of China. One of the first acts of the new regime was the establishment of the Chinese Film Bureau in 1950. Regulations were issued requiring all film, animation and documentary scripts to be cleared with the newly established bureau, and a censorship committee was set up to decide whether films submitted to the Bureau were suitable for production (Zhang 2004: 190). The Film Bureau also assumed responsibility for deciding which foreign films could be shown within China. When China's film industry was formally nationalized in 1952 the CPC assumed direct control over the administration, distribution, exhibition, production and criticism of all films made within the PRC. According to Zhang Yingjin, author of *Chinese National Cinema*, in taking these steps: 'The CCP sent the signal to filmmakers that cinema was no longer a simple matter of business or art, but rather a serious political operation subject to strict censorship from start to finish' (Zhang 2004: 191).

During the Great Proletarian Cultural Revolution, which lasted from 1966 to 1976, China's film industry ground to a halt. Universities and film schools were closed, no films were made between 1966 and 1972, and those that were made between 1973 and 1976 reflected the radical political agenda that dominated cultural activities in China during this period (Jia 1998). Following the Cultural Revolution, cinema once again became a popular form of entertainment. Although the sector remained heavily regulated and censorship guidelines continued to play an important role, numbers of annual feature productions increased steadily, from 67 in 1979 to 151 in 1986 (Zhang 2004: 227).

The early 1980s saw the emergence of the so-called 'fifth generation' – the first group of filmmakers to have graduated from the newly reopened film schools (Zhang 2004: 227). Although film remained a state-controlled industry throughout the 1980s, at various times studios enjoyed relatively high levels of freedom and discretion when choosing projects. Visionary studio heads were able to direct funding to the experimental, artistically driven and often commercially disastrous projects that came to characterize the emerging 'fifth generation' – directors such as Chen Kaige, Zhang Yimou and Tian Zhuangzhuang – many of whom also worked as 'underground' filmmakers at times. State-owned studios were able to support productions such as Tian Zhuangzhuang's *On the Hunting Ground*, which sold only two prints, for archival collection rather than general distribution, and *The Horse Thief*, which sold only seven prints (Zhang 2004).

REFORM AND OPENING UP

The pre-reform studio system offered filmmakers life-long job security, a fixed salary, government-subsidized housing and free medical care – the 'iron rice-bowl' of socialism. Although filmmakers working within this system had only limited freedom to choose their projects, they did not have to contend with the demands of a commercially driven entertainment sector. As Sun Shaoyi has observed:

> Because film was then largely regarded as a propaganda tool rather than a medium for artistic expression or entertainment, issues like film financing and marketing were of little importance for Chinese filmmakers. To a large extent, they were ideologically constrained, artistically restricted, yet, ironically, commercially worriless. (Sun 2000)

However, by the late 1980s, economic reforms and the growing availability of alternative forms of entertainment such as television, karaoke bars and live concerts combined with changing social patterns to alter the position of studios within China's entertainment landscape. Changes to the system of funding film productions forced studios to pay greater attention to the commercial viability of projects. Under the planned economic system the Film Bureau issued quotas for feature film production, and studios received a flat fee of RMB700,000 from the China Film Corporation regardless of box office takings. This changed when new reform measures were put into place in the 1980s. Distributors now either paid RMB9,000 per print or split the revenue with the studio (Zhang 2004: 239).

In 1994 *The Fugitive* became the first foreign film to be screened in the PRC on a revenue-sharing basis. Informal domestic quotas were applied to foreign films, stipulating that no more than ten films per year could be imported from outside the Chinese mainland (Lancaster 2001). When China joined the WTO in 2001 this number was increased to 20 films per year. In spite of tightly controlled limits over the number of foreign films that could be legitimately screened by Chinese cinemas, 80 per cent of China's box office revenue in 2003 came from just 20 imported blockbusters (Hua 2004). This represented only a 10 per cent increase in box office revenues generated by foreign films prior to WTO entry (Lancaster 2001). Nonetheless, the film industry's dependence on foreign films to generate revenue provided a stark reminder of the limited capacity of the Chinese film industry to satisfy the demands of Chinese audiences.

CHANGE AND ADAPTATION

By 2004/05 film production in China had been the subject of continuing processes of market-oriented reform for more than ten years. The role of film-makers was changing, and older systems of direct state control over production were gradually being replaced by new frameworks intended to ensure that the state maintained a high level of control over film content in the context of increasingly commercial approaches to film financing and exhibition. Spaces in which entrepreneurial modes of interaction could be applied to film were gradually being formalized and, although unauthorized copying and distribution in the form of 'pirated' DVDs was widespread, the role of intellectual property law in film financing and production in China was increasing. There were signs that some Chinese filmmakers were actively seeking out opportunities to engage with the market. Others were struggling to deal with the shift from a state-funded system to one in which a commercial focus was required, and had little understanding of how copyright functioned or the relationship it might have to their work (Wang and Wang 2004).

From a market perspective, film production in China could be broadly divided into four categories: legitimate mainland films, underground mainland films, co-productions and foreign imports. From a legal or administrative perspective these four categories could be further simplified into just two types of filmmaking activity: legal and illegal, legitimate and underground:

- *Legitimate industry – within the scope of government regulation:*
  - State-owned studios
  - Hong Kong films (exempt from quota restrictions)
  - Foreign co-productions
  - Quota imports
  - Independent productions
- *Underground film – outside the scope of government regulation:*
  - Films made or distributed without government approval
  - Films generally tolerated but not encouraged by the state
  - Films that often find international distribution
  - Films distributed illegally within China through pirate DVD outlets

In reality, filmmakers often moved between categories – perhaps honing their production skills making 'underground' films before finding commercial backers and moving on to 'legitimate' productions, or working on approved productions most of the time but occasionally choosing not to apply for government licences for specific projects.

Films made in China by foreign studios for both international and domestic audiences were also emerging as a significant component of the industry. The number of co-productions between Hong Kong and the mainland, in particular, was increasing rapidly (Law 2004). The Closer Economic Partnership Arrangement (CEPA) between Hong Kong and Beijing, which came into effect in 2004, allowed Hong Kong films to bypass the mainland's 20-film import quota. As long as Hong Kong studios employed sufficient numbers of mainland staff in their productions their films could be classified as Chinese productions rather than foreign imports (Law 2004). However, all co-productions required pre-production approval from the State Administration of Radio, Film and Television's Film Bureau. They also required post-production approval from the same authority before they could be distributed within the mainland. Furthermore, the Film Bureau stipulated that in order to be submitted to foreign film festivals a film must have been approved for cinema distribution by SARFT (State Administration of Radio, Film and Television 2003). As a result, regardless of who owned overseas copyright, the Chinese government maintained the power to decide whether a movie could participate in foreign festivals.

In 2001 quotas for foreign films imported on a revenue-sharing basis had been increased from 10 to 20 films per year in accordance with commitments made to secure World Trade Organization membership (CMM Intelligence 2004). Film was also required to compete with a growing range of entertainment options, including widely available 'pirated' DVDs and increasing access to content through the internet. Wholly state-funded studio productions had become the exception rather than the rule: in 2004 only 30 of China's 212 films were made by state-owned studios. The rest were made with private and/or foreign investment (Hui 2006: 63). In spite of a reduction in state funding for film production, China's mainland film industry produced 260 films in 2005 (China Film Press 2006). This figure was almost double the 140 films made in 2003 and represented a remarkable comeback for an industry that produced just 71 feature films in 2001 (China Film Press 2002). (See Figure 3.1.)

As *The Economist* reported, China's 2005 tally of 260 feature films was topped only by the United States, which produced 425 films, and India, which produced 800 (*Economist* 2006). Box office receipts were also growing: box office takings totalled 2 billion yuan in 2005, an increase of one-third compared with the year before (*Economist* 2006).

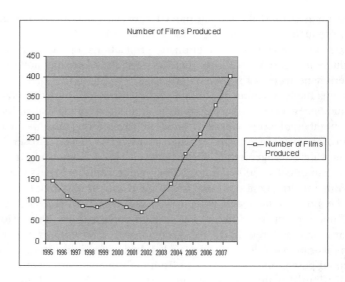

Sources: China Film Press (1995–2006); European Audiovisual Observatory (2009).

Figure 3.1   PRC annual feature film production 1995–2007

The Centre for Cultural Policy Research's *Study on the Relationship between Hong Kong's Cultural and Creative Industries and the Pearl River Delta* attributed the revival in Chinese cinema largely to the industry's liberalization and reform since the early 1990s (Hui 2006: 64). Wan and Kraus described this process as one of 'full-scale commercialization along with a vertical integration of production, distribution and exhibition sectors, and a linkage of film and television interests' (Wan and Kraus 2002: 420).

However, according to Zhang Hongseng, deputy director of the State Administration of Radio, Film and Television's Film Bureau, in 2005 only 90 of the 260 films produced by the mainland film industry were ever screened within China (Sauvé *et al.* 2006: 21). This admission suggests that increased volume of film production did not tell the whole story. Many of the films that did make it to the cinemas were withdrawn a few days after screenings began because of a lack of interest. Between 1990 and 2004 average per-capita annual income in China had grown from 1516 yuan (US$190) to 10 128 yuan (US$1271). This represented a sixfold increase in average disposable income (National Bureau of Statistics of China 2005: 337). In spite of these figures, film admission revenue had declined from RMB2.36 billion in 1991 to RMB0.9 billion in 2002 (Zhang 2004: 282). In 2005 China registered 200 million cinema attendances. Although this represented an impressive recovery from the lows of the late 1990s, for a population of 1.3 billion the figure was still relatively small (*Economist* 2006).

There were a number of reasons for this. Economic reforms had been associated with a sharp decline in the number of cinemas operating across China. State-sponsored mobile projection units that had serviced rural areas under Mao had also been abolished, resulting in the loss of a key point of access to cinema exhibition for many of China's residents (Jia 1998: 12–13). State-owned enterprises or unions, which had previously purchased tickets in block bookings to hand out to employees, also came under financial pressure, and the number of tickets sold in group bookings declined from 70 per cent in 1980 to 20 per cent in 1998 (Wan and Kraus 2002: 422). During the same period, the range of entertainment options for ordinary citizens increased dramatically.

In contrast to other genres of creative production in China, such as music, book or magazine publishing, film and television remained subject to strict pre-production censorship (Wan and Kraus 2002). China also lacked a film rating system – an important tool used to match audiences with content in other markets.[1] Without such a system, films approved for release were required to avoid content or themes considered unsuitable for any members of an audience, including unsupervised children. But, while the Chinese government was holding firm in relation to content guidelines for films screened and sold legally in China, access to less politically conscious entertainment products continued to expand. Although official distribution channels such as cinemas and television networks were able to screen only approved films, well-established illegal distribution networks, cheap home viewing equipment and broadband connections meant that consumers had ready access to material that had not been officially vetted.

Because unauthorized distribution, by definition, takes place outside the regulatory frameworks provided by the state, these forms of distribution avoided the bureaucratic web of import quotas, censorship requirements and distribution bottlenecks that plagued the rest of the industry. Films distributed through official channels, the most financially significant of which was cinema, were required to comply with the government's views on appropriate content, while at the same time satisfying the demands of audiences and investors. In contrast, unauthorized distribution networks were able to provide the films demanded by audiences, whether that meant Hollywood blockbusters, Korean romances, Japanese animation or the latest offerings of Chinese directors working either within the approved system or making 'underground' films, without complying with official production regulations or contributing to the costs of film production.

INSIDER VIEWS: PRODUCTION AND DISTRIBUTION BOTTLENECKS

This section of the chapter draws on interviews with Chinese filmmakers working in China in 2004 and 2005. Although all of the filmmakers interviewed were asked specifically about their views on copyright and its role in China's film industry, none saw intellectual property protection as the sector's biggest challenge. Access to revenue-generating distribution opportunities within China was viewed as the biggest problem for some directors, while for others the overly restrictive nature of the censorship system was a major source of frustration. For all of the directors interviewed, an underdeveloped cinema system, high ticket prices and difficulties associated with securing the funding needed to make films were seen as important factors in the industry's struggle to develop commercially.

The growth of unauthorized film distribution in China might be understood as the predictable reaction of a market in which access to the means of copying, distributing and consuming film has expanded rapidly, while tight political control and a cumbersome bureaucracy has continued to limit the availability of content through legitimate channels. 'Underground films', films that are made without government permits or approval and which cannot be legally distributed within China, might be framed as a similar development at the production end of the industry. The equipment needed to make films was once available to just a few filmmakers working for state-owned studios. Changes in technology have increased access to filmmaking equipment dramatically, and the privatization of large sections of production has resulted in new opportunities for filmmakers to work both within and outside the official film production system.

However, increased access to opportunities to make films has not been matched by a growth in access to coordinated distribution channels. Wulan Tana, studio-based director of *Warm Spring* (2002) and winner of the 2003 Golden Rooster Award for best directorial debut, suggested that underground films which cannot be distributed within China legally, but have proven popular among international audiences, are in many respects a natural response to limited distribution opportunities for filmmakers inside mainland China:

> More than 200 films are made each year and only about one-tenth of them can be shown in the cinema. The remaining 180 films have to look for overseas markets to survive. It is inevitable and it will promote overseas distribution. It is a reasonable result of a Chinese film industry which is under heavy pressure from Hollywood films. The appearance of underground films is necessary. Some underground films have given up the domestic film market and audience. They are sent abroad to win prizes. I think if these films could get enough recognition, they would have not been made underground. I believe that every director wants their films to be seen by people of their own country. (Wulan 2004)

Although it is clear from the outset that 'underground' films will not have access to cinema distribution inside China, Wulan Tana estimated that just 10 per cent of films made with full approval were ever released in cinemas. Some, such as *Where Have All the Flowers Gone?*, went directly to DVD (Ruggieri 2002). Others, such as Chen Daming's *Manhole*, took their chances in the highly competitive international distribution market. As a result, for many filmmakers, the time-consuming process of applying for pre- and post-production approval and the demands of tailoring content to the tastes of censorship committees offered few benefits.

Wulan Tana did not see a link between China's censorship system and the difficulties associated with finding distribution opportunities. In her words:

> Nowadays, China Film Bureau has a very open policy. There are no restrictions on the themes of films in China. So long as your films are acceptable for audiences and are not too far beyond the moral standard, you can make any kind of film. Chinese directors are not asked to make political films and would never be banned from making films unless their films contain too much unhealthy or negative content. (Wulan 2004)

However, Wulan Tana was speaking from a privileged position as a member of an established filmmaking elite who enjoyed access to studio support and funding, and who had successfully navigated cinema distribution within China. Although she also identified the cost of cinema entry and the number of cinemas available to audiences as important, she believed that the gravest concern for Chinese filmmakers was competition from big-budget foreign imports. In her view, most Chinese directors did not have access to the investment necessary to compete successfully with foreign blockbusters:

> it is unfair for a film with low investment to compete with one with much higher investment. Few Chinese directors can obtain the investment of RMB200 million as Zhang Yimou does. Zhang Yimou and Chen Kaige are quite exceptional in Chinese film circles. Most Chinese directors are in a situation similar to mine. We are not afraid of competing with those on the same budgets. (Wulan 2004)

Wulan Tana appeared to have accepted the existence of the pirate DVD market as a fact of life which she had no power to change. Furthermore, perhaps tellingly, her films had proven to be unpopular with the pirates. The most likely reason for this is that pirates adhere to commercial imperatives and stock films that sell well and are popular with audiences, rather than films which convey positive moral messages and are ideologically correct.

Li Yang trained at the Beijing Film Institute during the mid-1980s, before moving to Germany, where he worked as an actor and studied filmmaking at the Academy of Media Arts in Cologne. Li Yang viewed the government's role in the film industry very differently from Wulan Tana: he felt that the censorship system was unpredictable, impractical and unfair. His film *Blind Shaft* [Mang Jing] (2003) won awards at the Berlin, Hong Kong and Buenos Aires film festivals. *Blind Shaft* was filmed illegally in Northern China's coal mines and tells a dark story of two miners who conspire to murder a young boy in order to claim compensation for his death by posing as his relatives. The film, which is implicitly critical of conditions in China's coal mines and explores the desperation and poverty of those forced to work in them, was not approved for release in China.

According to Li Yang, the problems caused by censorship were being compounded by the lack of a law pertaining specifically to film. China's film industry is governed largely by administrative regulations, and a high level of discretion exists within opaque bodies such as the censorship board. According to Li Yang, the opaque nature of the censorship system and the film board's narrow views about what constitutes suitable subject matter for Chinese audiences, rather than commercial considerations, forced many Chinese filmmakers to find other ways of making their films. In his words:

> The underground films in China are made because of the system of censorship. The censorship system is very ridiculous, for there is no standard. They say, 'Your films are illegal', and then I ask, 'What is the relevant film law? Which clause do my films violate?' No, there is no specific law. So the censorship is quite cruel. If they say your film is 'illegal', then it is 'illegal'. (Li 2004)

In Li Yang's view, the censors' insistence that filmmakers create only positive images of China was stifling artistic expression and driving talented individuals away from the legitimate film industry, and often away from China. Li Yang believed that many filmmakers had been forced to make their films underground because they were unwilling to comply with the government's demands that they portray only wholesome, positive images of life in China:

> You cannot make a film in which there is a love triangle. Then if you make a film about ordinary people's life in small lanes, they would say you are smearing the image of China because the streets in the films are so dirty. Then what else can you do? (Li 2004)

Furthermore, although industry reforms had succeeded in commercializing some areas of filmmaking, expensive propaganda productions which were virtually guaranteed to be box office flops were still being made. Li Yang pointed to the release of the 2004 film *My Days in France*, directed by Zhai Junjie, which follows Deng Xiaoping's time in Paris as a young student, as stunning proof that this type of policy continued to play a major role in the allocation of valuable government funding for production: 'Can you imagine who will watch it? Definitely not many. It is a waste of money' (Li 2004).

However, according to Hu Bo, who trained at the New York Film School and was the director in charge of film productions surrounding the 2008 Beijing Olympics, *Blind Shaft* was banned by the authorities because it failed to present China in a positive light. In Hu Bo's words:

> The makers of those films purposely try to sabotage the ideology. That kind of film was sure to be censored. Their vision, their strategy is very clear. They try to show the dark side of the current society. Obviously, they achieved their creative vision. They won. But although it is very creative the government is 100 per cent justified in censoring it. Because they felt it would hurt China's image. I think that's their reason. . . . The film is very creative. . . . I like the film in an artistic way, but as an Olympic film producer I don't think it would be proper to use this film as a national image. (Hu 2004)

Hu Bo agreed wholeheartedly with the government's view that one of the primary obligations of Chinese cinema is the promotion of 'positive images of China'. Hu Bo sympathized with the artistic desires of underground directors such as Li Yang. However, like Wulan Tana, Hu Bo did not believe that censorship was the cause of the industry's struggle to produce commercial successes. Rather, censorship was just one factor in the highly complex environment of finance and regulation that directors and producers had to navigate to get their pictures made. Hu Bo pointed out that directors of mainstream productions seldom enjoy complete artistic freedom or autonomy in any film market. Even in Hollywood, scripts are altered or rejected because they don't match formulas. Although this is certainly true, one key difference between the vetting of scripts by Hollywood studios keen to produce a box office smash and Chinese censorship boards is that audience taste, as it is measured by profit, drives decisions in the Hollywood system.

In any case, like Wulan Tana, Hu Bo didn't believe that censorship was having a negative impact on audience perceptions of Chinese films. He argued that audiences, either consciously or unconsciously, had developed tastes which coincided with those of the censors. Hu believed that Chinese audiences had grown used to watching films that told a certain kind of story in a specific way, ensuring that they did not notice the impact of censorship in the same way that foreign audiences might. In his words: 'Right now, I don't think people are very attentive to the censorship situation, but they consciously or unconsciously like to look at things this way because it has become a habit' (Hu 2004).

Hu Bo's view that Chinese audiences are not alive to the impact of censorship on film was not shared by Chen Daming, who argued that the sophistication of Chinese audiences is often underestimated. In 2004 Chen Daming was a relatively young director who had trained in the US and returned to China with the explicit goal of working within the limits of the legitimate system.

According to Chen Daming, one of the benefits of unauthorized distribution has been that ordinary Chinese consumers have had the chance to develop both the habit of watching films and increasingly refined tastes. He believed that high-quality European films are much more popular with ordinary Chinese than mainstream 'Hollywood trash' (Chen 2004). However, in 2004, determining the tastes of Chinese audiences remained a highly subjective process. In contrast to the cut-throat, profit-driven focus of the US film sector, which researches and maps the tastes of audiences in order to minimize the risks associated with investing in high-budget productions, high rates of pirated DVD consumption and relatively low levels of cinema attendance made it difficult for China's filmmakers to find the most basic statistics on audience tastes and consumption habits.

According to Chen Daming, while underground filmmakers were not required to modify their content for reasons of official censorship, their productions did not necessarily paint a less distorted image of China:

> China's underground film industry is very dynamic. There are so many changes taking place in China. You probably couldn't find another country undergoing such a total process of change. Underground filmmaking is also popular with investors. However, this doesn't save the film industry. Every kind of underground film can win an award – film festivals are crazy about underground films. Audiences like watching underground films because they can be watched with a surrounding story – they are made in the context of questions about persecution, mystery and political criticism. But they paint a stereotyped picture of China. They are not a true reflection of China. (Chen 2004)

Chen Daming's 2002 film *Manhole* [Jingai] had to be substantially rewritten and renamed before it was passed by the censorship board. Once Chen Daming had dealt with those hurdles, he still faced the challenge of finding a distributor willing to take the film on. According to Chen Daming there are two kinds of distributors: mainstream distributors such as Sony Picture Classics, which are part of the Hollywood system and interested in mainstream epic productions, and smaller distributors, which are interested in edgy, often underground films. His film is neither of these and so, he believed, continued to languish.

In spite of this, Chen Daming remained firmly committed to his goal of making above-ground films in China for Chinese audiences. He believed that what the industry needed was high-quality cinemas, an electronic ticketing system, better promotion, genuine attempts to commercialize the DVD market and a wider range of films for audiences to enjoy:

> In Australia and the US multiplexes show maybe 30 different films. DVDs in China provide a massive range. Movie theatres only offer two or three choices. If China fixes its theatre system and provides a bigger range, Chinese audiences will come back. Chinese audiences love movies. (Chen 2004)

Simon Lan, film producer for the private production company Beijing Frontline Productions, highlighted the role of both practical and artistic factors in deciding whether or not to apply for approval to make and release films within China. Lan graduated from the Beijing Film Academy (BFA) in 1993. In 1998 he and a group of fellow BFA graduates began producing television commercials in order to make a living. By 2000 he and his colleagues had begun to establish themselves financially and decided to return to their first love: feature films. In 2005 Beijing Frontline Productions was making between 24 and 30 commercials per year (Lan 2005), and the company was using the income generated by television commercials to fund feature film productions. By 2005 they had made three films: *Incense* [Xiang Huo] (2002), *Soap Opera* [Feizao Ju] (2004) and *Mongolian Ping-Pong* [Lu Caodi] (2005). The group chose to make *Soap Opera* as an underground production, treating it as an opportunity to indulge their film-making passions and to experiment with the idea that they might go on to make commercially appealing productions in the future (Lan 2005).

Given its tiny budget and relatively inexperienced production team, *Soap Opera* was enormously successful. It won the title of 'Best New Asian Production' at the Pusan International Film Festival, and Wu Ershan received the 'Best Director' award at the Geneva International Film Festival. Simon Lan is adamant that the group did not decide to make *Soap Opera* as an underground production in order to ensure that it would be more appealing to international film festival judging committees. Rather, it was a practical decision: dealing with censorship is a time-consuming and often frustrating process. Although he thought that *Soap Opera* probably would have been approved eventually, it was simply not worth applying, because the group had no desire to release the film commercially in China (Lan 2005).

Hu Bo, Chen Daming, Li Yang and Wu Lantana all identified China's film distribution system as a major sticking point for the development of the film industry. In 2004 China had only one cinema for every 122 000 people, compared to the US, where there was a cinema for every 8600 people (Miller *et al.* 2005: 323). Total box office revenues in China were only US$120 million in 2003, compared to US box office revenues of US$9.5 billion (*China Economic Review* 2004). According to Li Yang, the popularity of pirated DVDs was a result of a cinema system that was out of tune with the market:

> Chinese films do not lack a market because of piracy. Everybody knows the quality of pirated DVDs is not good, but why don't people go to the cinema? Because the price of tickets is too high: about 50–80 yuan. That is almost the same as the price of cinema entry in America or Australia. Chinese people's salary is only one-tenth of American people's salary. (Li 2004)

Chen Daming believed that, when Chinese audiences were given access to multiplexes and cinemas were able to offer a better range of films, audiences would return. As he pointed out, in markets such as the United States, going to the cinema is about an experience – a night out, followed by dinner or a café. However, in 2004 there was little to draw audiences away from pirate DVD shops offering literally hundreds of titles for very low prices, into cinemas which might offer just two or three titles at comparatively high prices. In the words of Chen Daming: 'DVD stores are not about price. They are about selection. If you go to a DVD store you can choose from hundreds of DVDs. If you go to a cinema you can choose from just two or three titles' (Chen 2004).

Hu Bo believed that a lack of opportunities to express entrepreneurial ambitions within the formal film distribution system meant that the drive for profit was being expressed by vendors of pirated DVDs (Hu 2004). Perhaps surprisingly, Li Yang, whose films could not be distributed legally within China, also identified the pirate DVD industry as one wholly driven by profit, and not one which he supported:

> although many people understand the idea of 'copyright', and they know piracy is illegal, they still do it because they want to earn money, just like the drug addicts. It has nothing to do with poverty; it is out of the desire of making profits. (Li 2004)

SPACES FOR COPYRIGHT?

As discussed earlier, until quite recently Chinese cinema existed in the context of a media wholly dominated by the state. Films were produced by state-run studios and were understood primarily as an educational tool rather than as a creative or commercial medium. Although the Chinese consumers of the twenty-first century have access to a very wide range of content produced both within and outside the supervised production system, the Chinese government continues to regard film's educational function as a key concern in the formulation of policies related to the industry. According to director of the Chinese Culture Promotion Society, Wang Shi:

> In the past, the Chinese government regarded culture as an educational method through which people learned to be patriotic and upright, or as a kind of life or entertainment. This is a tradition. Even in ancient China, ideologists thought that culture was an important path leading to ethical or political education. However, things have changed since reform and opening up. Now, culture in China bears two responsibilities. One is the function of education; another is its function as an industry. I think the government puts educational function at the first place, but its function as an industry makes it regard market and profits as the primary concern. (Wang 2004)

The new commercial imperatives faced by China's film industry have given rise to the need for a coordinated system of distribution and revenue collection capable of providing incentives for private investment in film production. In other markets intellectual property rights have played an important role in the development of film industries, linking consumption with a return on investment. However, China's intellectual property system was not only neglected under communism, but the idea that individuals or companies would own an exclusive right to a creative product and would dictate how that product was used was in many cases viewed as a notion that conflicted directly with the goals of the state. Media works created by state-owned bodies were considered common property (Qu 2002), and studios themselves saw little or no need to 'protect' their products from unauthorized use.

Both revenues and audiences were guaranteed in the sense that the state was the only purchaser of film products and the only source of industry finance. Films were made in order to provide audiences with ideological and moral lessons in an entertaining context, to motivate and to uplift the audience, to stir patriotic feelings and to encourage pride in the achievements of the nation. As long as the state was responsible for film financing and distribution, intellectual property rights had little role to play. Rather than focusing on protecting the rights of studios and preventing unauthorized distribution, the system was designed to ensure that films conveyed messages consistent with the goals of the state and that the widest number of people possible viewed such productions.

The rejection of the idea that an individual could own intellectual property rights or deserved to be acknowledged as an author or artist by the CPC fitted comfortably with many of the Confucian notions of learning and creativity, as well as imperial approaches to publishing, discussed by Alford (1995). Restrictions surrounding publication or dissemination of works, including film, were aimed at suppressing dissident views and ensuring that works that were made available to the public accorded with the needs of the state. The commercial ambitions of cultural producers had very little role to play in such a system. As Mertha (2005) observes, even after the formal enactment of a copyright law in 1990, protecting the rights of newly recognized copyright owners has remained low on the nation's list of priorities, as just one issue among many to be dealt with by an overstretched bureaucracy and legal system (Mertha 2005: 226–31).

Copyright protection is often justified in terms of its ability to promote the creation of new works and to encourage economic growth. Nonetheless, the best way to formulate a copyright regime is a subject of ongoing and often heated debate (Halbert 1999: 28–36). Striking a balance between creating an economic incentive for the creation and dissemination of new works and restricting access to works by the community is at the centre of arguments about appropriate levels of copyright protection (Drahos and Braithwaite 2002: 3–5). Political economists emphasize the expansionary logic of capital as a driving force in the formulation of intellectual property policies in market-based systems. According to Ronald Bettig (1996):

> when it comes to the domains of information and culture, the logic of capital drives an unending appropriation of whatever tangible forms of intellectual property and artistic creativity people may come up with, as long as this creativity can be embodied in tangible forms, claimed as intellectual property and bought in the market-place. (Bettig 1996: 34)

Closer integration into a global economic system in which the value of 'filmed entertainment' was worth US$81 billion in 2006 has resulted in sustained international pressure on China both to increase levels of copyright protection and to allow greater international access to China's film market (the WTO case). As the role of the market in the film industry grows, commercial classes with an interest in protecting and expanding their entitlements are becoming apparent within China. Private investment in filmmaking, from both Chinese and international sources, has ensured that the ability to assign copyright has become a key tool for those eager to secure funding.

Bettig observes the role that copyright, when combined with an ability to control distribution systems, plays in making it possible for 'capitalist classes' to extract value from artistic and intellectual labour:

> to get 'published', in the broad sense, actual creators must transfer their rights to ownership in their work to those who have the means of disseminating it. With ownership of the means of communication and the exclusive control over the media product conferred by copyright, capitalists decide when and where to distribute artistic or literary works to achieve the highest possible return on their investments. (Bettig 1996: 35)

Although copyright is an important element of the cultural industries operating within capitalist economic systems, the commercial benefit that it confers depends heavily on the ability to access and control distribution. Unauthorized distribution in the form of 'piracy' is thus perceived as a major threat to industries such as film, because it disrupts controlled systems of distribution which allow copyright owners to dictate the terms on which products will be used or sold. High levels of unauthorized distribution in China have thus been viewed by international film industry groups as a major obstacle to realizing the economic potential of their products in those markets. According to Wang Shi in 2004:

> About 90 percent of disks sold in the Chinese market are pirated disks . . . people buy pirated products not only because of their economic capability, but also out of

their vague consciousness of copyright. Maybe all people know that piracy will harm those practitioners (writers, composers or film directors), but they may also think that this harm is tiny to those wealthy practitioners. Since the benefits are so great, almost nobody can resist the temptation. (Wang 2004)

In markets where levels of copyright protection are high and commercial organizations rather than government policy control access to distribution channels, film corporations act as mediators between creative labourers and the market. Copyright allows relationships between investors, producers, distributors and consumers to be managed. However, in the case of China, the government's policies of maintaining direct control over access to distribution systems and limiting the power of private investors through mechanisms such as censorship, import quota restrictions and heavy regulation over investment in the physical distribution infrastructure are restricting the role that both copyright and the market are able to play in the development of China's film industry.

ATTRACTING CONSUMERS BACK INTO CONTROLLED SPACES

While unauthorized distribution in China has undoubtedly been a boon for consumers, it also represents a significant reduction in the central government's ability to physically control content that is available to audiences. Attracting filmmakers and audiences back into controlled spaces of production and consumption, rather than attempting to remove physical access to unauthorized content altogether, might be understood as an important strategy for disciplining the film industry in the context of a globally connected population and a market-driven economy. As Wan and Kraus (2002) put it:

Like filmmakers, the party is concerned with how to implement market reforms in cinema. Unlike filmmakers, it is also attempting to carve out a new road for its propaganda. Its dilemma is how to commercialise propaganda to have an impact when consumers have unprecedented choice. (Wan and Kraus 2002: 428)

In 2009, there were signs that China's policymakers had succeeded in creating a policy environment that was allowing a profitable film industry to develop, while state control over content distributed through commercially valuable official distribution channels was being maintained. Huang Qunfei is general manager of the New Film Association, a company that operates 34 cinema multiplexes across China, with 497 screens. According to Huang Qunfei, between 2005 and 2008 box office revenue increased from RMB1 billion to RMB4.3 billion. The number of cinema screens had also increased from around 1000 to over 4000. Furthermore, whereas in the past just a few national film studios had been responsible for production, by 2009 the main investors in film were private companies (Huang Qunfei 2009).

Huang Qunfei was adamant that an expansion of spaces in which private investment and market-driven competition were able to operate had allowed the industry to expand rapidly between 2005 and 2009. As he put it:

> The reason for the industry's rapid growth has been that official policy has supported growth in the industry. For example, investment from private companies has been allowed to flow in. In the past, film companies were all government-owned corporations. After that, private companies were allowed to invest, but the process was very complicated: private investors had to purchase film production licences from government-owned studios, and films had to be released under the name of government-owned studios. Right now, if you have money, you then just apply for a production licence, and once you have approval you can produce a film. This is because policy has become less restrictive. Government policy on film is encouraging competition, the diversification of investment and the breaking up of monopolies. (Huang Qunfei 2009)

Although unauthorized distribution channels had not disappeared, and the number of broadband connections in China had continued to increase, Huang Qunfei believed that an increase in the number of high-quality cinemas available to Chinese consumers, combined with continued increases in standards of living, had resulted in decreasing rates of DVD piracy:

> the piracy rate is declining for several reasons. We have more and more modern multiplexes. People like to go to multiplexes to enjoy a new experience. In the past, our cinemas were old and weren't very attractive to young people. Now we have more theatres, and some of them are much better than American cinemas. Second, Chinese living standards have continued to improve. Seeing a movie in a theatre is totally different from watching a film at home. So people want to go to the cinema – they prefer going to the cinema to see blockbusters, because the experience is completely different from watching a pirated film at home. Third, some young people, fans of DVDs, don't like to buy pirated copies any more. They want to buy high-quality DVDs. So this is totally different from 2005. That means we have a bigger income as an industry. (Huang Qunfei 2009)

Access to films through film-streaming or downloading websites, which operate without permission from copyright owners, had expanded since 2005. However, product placement was providing an important source of production finance, and theatre distribution was producing significant investment returns for the industry:

> with Feng Xiao Gang's latest movie, they only invested RMB50 million, advertisements paid for more than half of production costs and box office revenue was RMB350 million. That's a 500 per cent return. TV rates provide very little money

compared with theatrical release. So the main revenue was from theatrical release. Also there is no merchandising in China, so the next step I think we will develop this kind of business, because the revenue from theatric release is stable. (Huang Qunfei 2009)

Huang Qunfei was jubilant about the Chinese film industry's prospects, and complained that his company's biggest concern was finding enough sites suitable for new cinema developments (Huang Qunfei 2009). Although the film industry was still expected to operate within frameworks on content set by the state, he did not believe that this would prevent the continued growth of China's film market:

Because China's understanding of the role of films is different from the West, this is our situation, and so far we haven't been able to change it. We must respect the government's regulations on the role of film. Within this system, we will try our best to do business. More domestic films will be produced, and the market will continue to grow. (Huang Qunfei 2009)

CONCLUSION

Wang Gungwu, director of the East Asian Institute at the University of Hong Kong, argues that, while China's government is still authoritarian, much of its power has been decentralized. The power of the Communist Party of China is now limited to key points that can still be controlled. While the CPC puts down mass protests, maintains close involvement in the press, detains political prisoners at will and attempts to control the internet, entrepreneurs and businesspeople are granted much higher levels of freedom. Individuals capable of adapting to the market system are being given the space and autonomy necessary to produce extraordinary levels of economic growth. By adopting this strategy, the Communist Party of China is succeeding where other socialist states have failed. The political system remains authoritarian, but the economic system is making the shift to a successful market economy (Wang, in Colvin 2006).

In 2004 and 2005 directors working in China were struggling to adapt to an increasingly competitive commercial environment in the face of the state's attempts to balance their desire to maintain control over the industry with the need to find commercially sustainable models of funding and production. By 2009 it appeared that, to a large extent, China's policymakers had succeeded in creating a policy framework that maximizes the benefits of market-driven investment and entrepreneurial activity in the film industry, but maintains state control over content. Although unauthorized production and distribution continue to exist, opportunities for profit within state-controlled spaces are expanding. Because the state has maintained a high level of control over cinema distribution, the only form of distribution capable of ensuring that popularity is translated into scalable profit for film industry investors, important commercial incentives for operating within state-controlled spaces have been created. Private investment is being mobilized to achieve what the state, on its own, could not: China's commercially focused film industry is finding ways to tell popular, entertaining stories consistent with the central government's views on the role of film as a tool of propaganda and pedagogy.

In an industry in which international investment, expertise and business models are increasingly important, copyright's role is expanding. However, the state's role in controlling access to commercially valuable distribution channels has, to some extent, usurped the power of copyright owners described by Bettig (1996). By limiting international access to China's film market, but allowing foreign investment and skill transfer, the Chinese government has also created space in which the domestic industry is able to develop a competitive advantage. These strategies might be understood as valuable tools in the struggle to create positive images of China that will not only appeal to Chinese audiences, but also be capable of competing in an international media-scape. A Chinese film industry that is capable of exporting approved films to international audiences would be highly valuable to China in commercial terms, as it attempts to shift away from low-value manufacturing industries towards higher-value-added areas of production. It would also offer China's policymakers an opportunity to shape perceptions of China in the eyes of global audiences: a valuable prize for a government that is eager to develop its global influence and expand its 'soft power'.

NOTE

1. This was still the case as of 2009.

4. China's music industry: space to grow

INTRODUCTION

In 2006, the CEO of the International Federation of Phonographic Industries (IFPI), John Kennedy, addressed the International Forum on the Audio Visual Industry in Shanghai, stating: 'Illegal sales of music in China are valued by [the] IFPI at around US$400 million, with around 90 per cent of all recordings being illegal. No creative or knowledge-based industry can hope to survive in such an environment' (Kennedy 2006).

Recording industry lobbies such as the IFPI and the Recording Industry Association of America (RIAA) have been engaged in high-profile campaigns against copyright infringement in China since the early 1990s (Wang 2003). It would be a truism to state that the intellectual property businesses that these groups represent have been frustrated by their inability to function effectively in the face of China's weak copyright environment.

In spite of John Kennedy's lament that high rates of unauthorized copying and distribution make it impossible for businesses to survive, some creative and knowledge-based businesses *are* making money in China. While John Kennedy and the IFPI were earnestly calling on China's authorities to increase levels of enforcement in order to make it possible for a commercially focused music industry to develop, in 2006 the chairman of the Chinese Audio-Visual Society declared that the nation had 'leapfrogged into an age of digital music', noting that digital distribution was outperforming traditional record sales, generating profits of 3.6 billion yuan in 2005, compared to just 2.7 billion yuan in the 'best year' of physical music sales in China (Sun 2006). Sales of mobile phone music are quickly becoming the most significant source of music-related revenue (Yao 2007), and the cross-platform promotion of China's *Supergirl* series generated massive profits for the show's developers and mobile partners (Keane 2007: 122).

Drawing on interviews with members of the music industry in Beijing between 2005 and 2007, this chapter discusses the ways in which China's music industry is adapting to the technological and political environment within which it must function. The chapter briefly discusses key differences between the conditions that allowed an organized recording industry to develop in the West and those within which China's commercial music industry is

China's creative industries

evolving. It then explores the impact of cultural policies ostensibly intended to restrict the circulation of heterodox content on commercially focused production and exploitation of music. Although unauthorized content can be obtained with relative ease, censorship policies are providing Chinese content with significant advantages. Finally, the chapter examines the role of mobile distribution channels as a growing site of music-related investment and income.

Threats to the dominant position of record labels and the use of new technologies and business models to connect fans with music are not unique to China (Fine 2007; Hunt 2007). This chapter argues that China provides a valuable case study for researchers interested in the relationship between copyright and the growth of creative industries because, to a large extent, music-related business models in China are evolving in an environment where copying and distribution are difficult to control and very high levels of music 'piracy' are the norm.[1] Examining how and why commercial activities centred around the production, distribution and consumption of music are developing in the context of weak copyright and rapid adoption of new technology may provide important clues for researchers and policymakers attempting to understand how music industries adapt to changing regulatory and technological environments.

BACKGROUND: A BRIEF HISTORY OF CHINA'S MUSIC INDUSTRY

By the late 1960s music had become big business in the United States, bringing much higher rates of investment return than other areas of the entertainment industry (Lipsitz 1994). Pop stars, rock concerts, music festivals, commercial radio, music charts and, of course, record labels had become part of the social and economic fabric of the 'Western' world (Gronow and Saunio 1999: 135). Copyright and contract law, legal technologies that the recorded music industry relies heavily upon, were well established. Increasing affluence and innovations in physical technology, such as transistor radios and vinyl records, made the consumption of music more affordable than ever before. These technological developments combined with the growing affluence of post-war populations, an emerging consumer culture and a particular legal environment to create the conditions in which specific kinds of music businesses were able to flourish (Laing 2002).

In contrast, none of these conditions were present in the People's Republic of China at that time. The PRC, which formally came into existence in 1949, was caught in the midst of a radical communist experiment. The country was still desperately poor – more than 30 million people died during the famines following the Great Leap Forward in 1958, and in 1979 one in three peasants were reported to be living on the verge of starvation (Becker 2000: 24). While the United States and Western Europe operated as market-driven economies, China's economy was centrally planned (Ikels 1996: 3; Becker 2000: 297–8). During the Cultural Revolution (1966 to 1976) the state became even more closely involved in commissioning and producing cultural and entertainment products, which were seen as important pedagogic and propagandistic tools for the state (Kraus 2004). By the time the post-Mao leadership began loosening restrictions on trade and interaction with the outside world in the early 1980s, the music industry that did exist in China, including the live, recorded and broadcast sectors, was dominated by state-funded cultural troupes dedicated to writing and performing a limited repertoire of propaganda songs (Kraus 2004: 9).

Although liberal market reforms and gradual loosening of state control over entertainment progressed throughout the 1980s, the formal legal system had been deliberately dismantled during the Cultural Revolution (Chen 1999: 40). As Alford (1995) discusses, China possessed no indigenous equivalent to concepts of 'intellectual property'. Notions of individual creativity and ownership of ideas were not easily reconciled with Confucian concepts of creativity, knowledge and learning (Alford 1995: 9). For these and various other reasons, China remained without a copyright law until 1990 (Qu 2002; Mertha 2005).[2] At the same time, economic reforms were giving individuals growing levels of disposable income and new choices about how it could be spent. Technologies that made copying, distributing and listening to music affordable on a mass scale combined with increased contact with the outside world to ensure greater access to foreign entertainment and cultural products for most of the population (Brace 1991: 45). However, censorship policies continued to heavily restrict cultural production and legitimate distribution (Brady 2006).

The creation of 'neighbouring rights' in the early twentieth century made it possible for recording businesses to own the copyright in sound recording they had commissioned (Laing 2002: 185). This development resulted in commercial opportunities for record labels: businesses willing to invest in the production and promotion of music that could be sold to a mass market. Because access to the equipment and expertise required to record music was limited, it was generally possible for an increasingly organized recording industry to monitor and control the production and distribution of music 'products' – physical copies of sound recordings (Gronow and Saunio 1999). The business models that dominated the global recorded music industry in the second half of the twentieth century were based around record labels providing artists with access to recording equipment, mass production and distribution channels, and marketing and promotion services, and remunerating them on a royalty basis.

Legally enforceable intellectual property rights and physical technologies that made *controlled* mass production and distribution of music possible

(Bettig 1996) were key factors in the rise of a handful of highly integrated, transnational music corporations that accounted for 90 per cent of gross sales of recorded music in 1994 (Burnett 1996: 2). Although developments in physical technology, such as cassette tapes and recorders, presented challenges to the industry's ability to control copying, these changes occurred after markets, industry structures, professional organizations and group collection infrastructures had become established. As a result, the industry was generally able to respond in a systematic way, and incremental developments in analogue technologies of copying did little to disrupt its overall structure (Frith 2004).

In China, on the other hand, technologies for mass reproduction and consumption of recorded music became available in the absence of copyright law, an organized domestic music industry, or clear legitimate channels for the distribution of most foreign content. These technologies also became available as China was making the transition from a planned economy to a market system. High levels of demand for popular music, combined with readily available technologies for mass reproduction and consumption and an absence of legitimate distribution channels, contributed significantly to the rise of a black market in music products and highly sophisticated illegal distribution networks (de Kloet 2002). The internet, personal computers and cheap MP3 players – technologies that are challenging approaches to the control and monetization of content globally – have compounded the difficulties associated with controlling distribution. Although there are some signs that the strategies of major international labels to online distribution in China may be changing, albeit slowly, the vast majority of music downloaded from the internet onto personal computers or portable devices such as MP3 players currently occurs without permission from or payment to copyright owners (Daniel 2007; International Federation of Phonographic Industries 2008; Music 2.0 2008).

Not only are new technologies being adopted with enormous speed across the country, but they are being embraced fastest by groups traditionally considered most likely to pay for music in other markets. Young, educated city-dwellers with relatively high disposable incomes are now the group most likely to have access to broadband connections, cheap MP3 players and next-generation mobile devices (CNNIC 2008). According to Kaiser Kuo, a founding member of the rock band Tang Dynasty and current director of digital strategy for advertising group Ogilvy:

> So many of the people who are interested in music are also internet users. If you look at 20- to 24-year-olds alone in China, the penetration of internet usage is over 40 per cent. It's about 43.4 per cent according to the last survey. And in urban areas, if you were to leave out rural China, I'm sure that it is upwards of 70 per cent. It's enormous. So all these people, the young urbanites in China are all internet users and everybody knows where to go to download music. It's no mystery at all. There's

no reason to go and buy any of it. So [record labels] have gone from the frying pan into the fire. They hadn't even put a dent in pirate physical copies before P2P and MP3 downloading came along. (Kuo 2007)

CULTURAL POLICIES

Commercial cultural and entertainment industries are a relatively new development in the PRC. Although content restrictions are being eased and commercial spaces are expanding, government approval is still required for the sale of music. Baranovitch (2003) argues that new technologies and a changing media environment have not necessarily diminished the state's interest in popular culture as a tool of propaganda. Rather, the Chinese government is adapting strategies of influence and control to function in an increasingly market-driven environment. By maintaining control over access to commercial opportunities, through control of the broadcast media, regulations requiring permits for large-scale concerts and the need to obtain publishing licences for legal sales of music, the Chinese government has been able to limit the commercial viability of artists it does not explicitly endorse (Baranovitch 2003: 271). Continued control over broadcast and print media also allows the authorities to prevent access to promotion opportunities for artists, businesses and products that offend political, ideological or moral sensitivities. Although it is still possible to make, play and listen to unauthorized music, the Chinese government has significantly limited commercial incentives for its organized production and distribution.

Policies originally intended to control heterodox content have had another important impact: creating barriers to the legitimate domestic market for foreign content producers, increasing incentives for the production of domestic content and reducing competition within the legitimate market for domestic music. In order to release an album, artists require a 'publishing number' that can only be obtained through a licensed publishing company. Foreign companies cannot obtain a publishing licence, and so they have no choice but to collaborate with a Chinese partner in order to distribute music. While licensed publishing companies are able to issue publishing numbers for local artists directly, foreign albums must be vetted and formally cleared for release by the Ministry of Culture – a time-consuming process. In 2005, Huayi Music's vice general manager Daniel Zhao cited this added difficulty as a factor in his label's decision not to release foreign albums:

It is very hard to release foreign albums in China. In order to get a publishing number you have to apply to the Ministry of Culture's censorship committee – song lyrics, album design all have to be approved by the censorship committee. It probably takes two to three weeks. Then, if there is no religion, no sex, no bad language, the Ministry of Culture may say OK. Sometimes they will say that certain tracks are not suitable to be released, so individual tracks have to be removed. That system doesn't apply to domestic albums. Domestic albums just deal with the book number system. Publishing companies themselves are able to issue the numbers. That process will only take four to five days. (Zhao 2005)

Beaker Huang, business development manager of Warner Music China, viewed the music censorship system as a straightforward market protection measure, intended to keep foreign artists out while the domestic industry developed (Huang 2005).

In *Bad Samaritans: The Myth of Free Trade and the Secret History of Capitalism*, Ha-Joon Chang (2008) argues convincingly that government intervention to protect infant industries from international competition has played a key role in the rapid economic growth and industrial development of countries that have established themselves as economic powerhouses. There are signs that a similar process is occurring in relation to China's music industry. Governing structures that came into existence to allow the state to maintain ideological control over culture are now shielding emerging creative and cultural industries from foreign competition. This is providing space within which audiences for home-grown content are being established and domestic content production capacity is being increased.

THE EVOLUTION OF A CHINESE APPROACH TO THE MUSIC INDUSTRY?

The dominance of a few developed nations in global trades in culture has led many to question the fairness of the expanding global intellectual property system. According to UNESCO, developing countries account for less than a 1 per cent share of exports of cultural goods (UNESCO 2005). It is well documented that developed nations have been a driving force in the expansion of copyright in an international context and that copyright industry lobbies based within these countries have influenced the international protection frameworks (Arup 2000; Wang 2003; Miller *et al.* 2005; Consumers International 2006: 4). Critics argue that the global intellectual property system does little more than create a larger market for the export of copyrighted content by nations in which these industries are well established. In addition to creating a danger that domestic cultural industries in developing economies will be unable to compete with a flood of foreign content, overly restrictive copyright laws prevent access to the information and materials that developing countries need in order to increase the education and skills of their own populations (Drahos and Braithwaite 2002; Consumers International 2006).

In spite of this, foreign hopefuls have been disappointed by their inability to capture a greater share of China's increasingly dynamic domestic music market. UNESCO reports that, in 2003, imported music accounted for over 50 per cent of recorded music sales in China (UNESCO 2005: 39). However, given that even the most conservative estimates suggest that 'piracy' makes up around 80 per cent of China's physical music market (Credit Suisse Equity Research 2005), this figure must be understood in the context of a much larger market, for which reliable statistics are difficult to find. Other sources have reported that in 2005 Western artists accounted for just 5 per cent of music sales, and regional Asian artists for a further 50 per cent (Scott 2005). Industry insiders estimate that international artists still account for less than 10 per cent of the overall digital revenue of major labels operating in China (Daniel 2007). According to the International Federation of Phonographic Industries (2008):

> China has potentially the largest online music-buying public in the world with as many broadband connections as the United States. Currently, however, more than 99 percent of all music files distributed in the country are pirate and China's total legitimate music market, at US$76 million, accounts for less than 1 percent of global recorded music sales. (International Federation of Phonographic Industries 2008)

Why have the experience and accumulated catalogues of major international record labels counted for so little in China? Part of the answer appears to be that international labels have built their business models around the notion that an artist can record music, which can be sold independently of the physical presence of the artist in any particular market. Ed Peto (2007), Beijing-based promoter and music consultant, observes: 'While the traditional record label model isn't exactly going through a golden age in the West, it never even had a golden age in the Middle Kingdom' (Peto 2007).

While major international labels have been unwilling to invest heavily in the promotion of international artists in a market where mass scale returns are difficult to secure, local artists and labels have been actively working to develop business strategies capable of generating income in spite of very high levels of piracy. One strategy for doing this has been to rely on personal appearances by artists, which cannot be replicated. As a result, there is less emphasis on producing popular albums and more emphasis on gaining popularity and profile through single hits that lead to lucrative product endorsement and live appearance or performance deals:

> In foreign countries there is a clear line between publishing deals and record deals. But over here you would often find a record label playing all of the different roles as a record company, as artist management, as a publisher. They sign all-round deals with their talent. They do make sure that once they invest in one field they recoup it from every form possible. (Wang 2005)

In contrast to Western markets, where artist management and music are generally separate, in China assigning a record label with management rights is considered one of the most important aspects of an artist's contract, forming a vital income source for domestic labels: 'We own both the music rights and the artist's performance rights. As a result, we get a certain percentage of the artist's performance fee every time they appear in public' (Zhao 2005).

Artists such as the Huayi-signed Yu Quan duo can earn up to 200 million yuan (US$25 million) a year in performance fees (Zhao 2005). However, even for Chinese labels, relying on personal appearance and advertising revenue presents practical problems. Personal appearances have limited scalability. Neither advertising nor personal appearance fits well with the 'long tail' approach, which, in other markets, allows back-catalogues to continue generating revenue for labels and artists long after the artist has been eclipsed by the latest trend.

It is difficult for international labels to capitalize on the established reputations of foreign artists by mimicking these practices in China. As with music sales, concerts by foreign artists are made more difficult as a result of complex licensing requirements for live events and vetting procedures relating to both artists and their music. The international music press widely reported the 17 July 2008 announcement by the Ministry of Culture that the political backgrounds of all foreign performers would be checked, and those considered a threat to China's sovereignty would not be granted permission to perform in China (Coonan 2008).

As China's media sector becomes more commercially focused, opportunities for collaboration, cross-promotion and integration are increasing. Chinese media groups are adapting formats and approaches developed elsewhere in the world with high levels of success, shielded from foreign competition by rules preventing foreign investment in the broadcast media (McCullagh 2005). In 2004 *Supergirl*, a Chinese equivalent of *American Idol*, earned Hunan Satellite television 68 million yuan. The Entertainment Package Company [Yule Baozhuang Gongsi], which was responsible for marketing the performers and aggregating ancillary rights associated with the programme, such as merchandising, concerts and CDs, accrued 78 million yuan in revenue (Keane 2007: 122). China's own capacity to capture commercial opportunities associated with music is increasing as media commercialization becomes more entrenched.

While China's domestic media industries have been developing the ability to make and sell content that appeals to Chinese audiences, exploring commercial opportunities associated with cross-platform promotion, growing mobile networks and access to the broadcast media, major international labels appear to have been caught in a negative cycle. It is extremely difficult to make money by licensing copyright in China, and gaining access to the market is expensive and difficult, so Western labels have devoted few resources to promoting their products to Chinese audiences. Mathew Daniel, vice-president of China's leading digital distribution company R2G, summed up the problem to record industry executives attending the 2008 MIDEM (music industry) conference:

> It has to be realized that the vast majority of labels at MIDEM are probably currently unscathed by piracy in China and that's likely because their music is so obscure in the Chinese consciousness that they have not even had the dubious honour of gracing the servers of China's notorious MP3 search engine, Baidu. Piracy in China often gets a lot of attention but many forget the other Ps of marketing and these are the basics that labels intending to come into China should first focus on. For dramatic effect, let me . . . quote Tim O'Reilly when he said that 'Obscurity is a far greater threat to authors and creative artists than piracy.' (Daniel 2007)

MOBILE DISTRIBUTION CHANNELS

According to the International Telecommunications Union, mobile penetration in China increased from just 1.9 per cent in 1998 to 30.3 per cent in 2005 (International Telecommunications Union 2008). The Ministry of Information and Industry reports that in February 2008 more than 56 million households had mobile access (Ministry of Information and Industry 2008). The distribution of music directly to mobile devices by mobile operators ensures that content use can be monitored. Mobile operators' established billing systems make it possible for revenue to be collected with few additional infrastructure or administration costs. The centralized nature of mobile communication networks also prevents many of the problems of unauthorized copying and distribution associated with the distribution of music through the internet, providing an effective digital rights management solution for copyright owners. In China, where unauthorized networks for the distribution of physical copies of music are well established, independent monitoring agencies do not exist and users have demonstrated low levels of willingness to pay a premium for 'legitimate' content, mobile music is emerging as one of the only large-scale points of distribution control and systematized revenue collection available to the market.

Mainland China has just two mobile operators: China Mobile and China Unicom (China Unicom 2008). With more than 369 million subscribers, China Mobile is the biggest mobile operator in the world, and the dominant player in China's mobile music market (China Mobile 2007). The company also boasts the world's most extensive mobile infrastructure, which extends to all of China's 31 provinces (Yao 2007). China Mobile Communications Corporation

(CMCC), a state-owned enterprise, is China Mobile's controlling shareholder, owning 74.33 per cent of China Mobile Limited shares as of 31 December 2007 (China Mobile 2008). In 2007, China Mobile reported 48 million wireless music club members, 22 million of whom were classed as 'senior members'. Thirty-one million of China Mobile's music club members had joined in the previous six months – an increase of nearly 65 per cent. China Mobile reports that 'colour ring revenue' has increased to 5.027 billion yuan in the first half of 2007, up more than 90 per cent on the same period in 2006 (China Mobile 2007). China Unicom is significantly smaller, with 'just' 142 million subscribers (China Unicom 2008).

This growth in mobile music subscribers within such a short period suggests that the distribution of music through mobile networks may mark a watershed moment for the commercial music industry. Kaiser Kuo, an original member of the Chinese heavy metal group Tang Dynasty, which is often credited with the title 'first heavy metal band in China', believes that: '[Labels] have a very different job here. Their job now needs to be entirely focused on digital and they should be looking toward full song mobile downloads' (Kuo 2007).

Just as analogue technologies allowed a limited number of firms in Europe and the United States to control the physical production and mass distribution of music for much of the twentieth century, mobile networks make it possible for a few key players to control distribution of music to mobile devices. In an environment where ownership of the copyrights associated with music has, until now, meant very little and domestic recording industry lobby groups have been unable to form, there is a very real possibility that mobile operators rather than record labels will emerge as the industry's dominant stakeholders.

The rate at which this new distribution method is being accepted by the public and the lack of alternatives for controlled mass distribution and centralized revenue collection for music copyright owners suggest that mobile music is likely to emerge as one of the most important sources of music-related income in China. As Bettig observes, in the context of transferable intellectual property rights, an ability to control the means of communication enables 'capitalist classes' to extract value from artistic and intellectual labour:

> to get 'published', in the broad sense, actual creators must transfer their rights to ownership in their work to those who have the means of disseminating it. With ownership of the means of communication and the exclusive control over the media product conferred by copyright, capitalists decide when and where to distribute artistic or literary works to achieve the highest possible return on their investments. (Bettig 1996: 35)

It appears that in China, where mobile operators control the increasingly important mass distribution channels, China Mobile is on track to play such a role. The development of the mobile music market in China has been led by ring tones and caller ring-back tones (CRBTs). While ring tones can be downloaded from a range of sources, CRBTs depend on centralized administration through a mobile network. Caller ring-back tones first appeared in Korea in 2002 and have rapidly become one of Asia's most popular mobile value-added services. CRBT services allow subscribers to choose a song that callers will hear in place of the standard 'ringing' tone when dialling the subscriber's number, and are widely understood to be at the forefront of wireless music services, introducing subscribers to the concept of music delivered directly to their mobile devices.

Unlike ring tones, which are generally stored on individual mobile phones, CRBT services are managed centrally, through mobile service providers. China Mobile launched China's first CRBT service in 2003 and now provides a nationwide CRBT network for all of China's 31 provinces (Yao 2007). By August 2006, China Mobile claimed to have 144 million CRBT users, with a penetration rate of 48.2 per cent (Yao 2007). China Mobile has reported staggering growth in revenue generated by ring tone and CRBT services (China Mobile 2007). The company's 2007 Interim Report states that colour ring revenue increased by more than 90 per cent in just 12 months, to 6.027 billion yuan (China Mobile 2007).

In addition to solving many of the problems associated with unauthorized copying and distribution of music, mobile distribution is allowing the industry to access groups of consumers who were previously considered beyond reach. China's mobile networks extend across China and are often available in rural areas, beyond the reach of music retailers, the internet and fixed-line services. In June 2007 China Mobile reported that half of all new subscribers came from rural areas (China Mobile 2007). Ring tones can be purchased for as little as 2 yuan (US$0.27) and CRBTs subscriptions for as little as 5 yuan per month (US$.69) (Daniel 2007), making them popular with lower-income groups that would be unlikely to purchase expensive legitimate copies of physical music. Top Chinese singers can sell between 10 million and 20 million CRBT subscriptions annually (Daniel 2007). As Mathew Daniel points out, 'this is a rarefied space that is not breached by Western artists'. Chinese artists are responsible for the vast majority of revenue being generated through mobile music (Daniel 2007).

Given the power of Chinese mobile operators to control distribution channels and revenue collection, it is unsurprising that the vast majority of money generated through mobile music in China stays with the mobile operator. Service providers, distributors, labels and music publishers share an initial 'sign-up fee' when a subscriber first signs on to a CRBT package. After the initial sign-up fee has been divided between these parties, subsequent full monthly subscription fees are kept solely by China Mobile (Daniel 2007).

While there are signs that the role of record labels in China will be very different from their role in other markets, Terry Tang, director of China business development at Noank Media (also known as Fei Liu Internet Technologies), suggests that they will not disappear from the market completely. While mobile music is quickly becoming the most important source of revenue in the industry, there is still a place for live music and artist management. Labels are in a position to connect artists with mobile operators, manage advertising deals and ensure that publishing and performance licences are obtained – all key components of the music business in today's China (Tang 2007).

The appearance of an accessible mass market for commercially produced music is prompting increased concentration and coordination of production activities among domestic operators. Although there have been some amazing stories of amateur musicians who have produced a hit ring tone from their bedroom and made millions, commercial spaces are quickly being filled by large, vertically integrated content providers. Hurray is one such company. In addition to providing mobile operators with the software applications designed to manage the delivery of mobile content services, Hurray has purchased record labels and artist management companies such as New Run (Birdman) and Huayi Brothers Music (Hurray 2008) and secured a partnership agreement with MTV China (China Tech News 2006). The continued significance of book numbers, a legacy of the censorship system, may also add to the concentration of power within the market of a few key players. It is interesting to note Hurray's purchase of Huayi Brothers Music, one of China's few record labels with a licence to issue its own book numbers. This suggests that Hurray has recognized the strategic advantage associated with the ability to issue these licences in an environment of growing demand for content.

CONCLUSION

Although much of the scholarly discussion of censorship in China has focused on its use in controlling the ideological nature of content, little attention has been paid to its role in protecting China's developing domestic creative and cultural industries from foreign competition. In the case of the music industry, it appears that cultural policies that make it harder to publish foreign content, strict regulations governing foreign investment in content industries, and low levels of copyright enforcement have worked together to provide Chinese media businesses with space to develop effectively. Although groups such as the International Federation of Phonographic Industries have focused heavily on the need to increase levels of copyright protection to ensure commercial developments in China's music industry, the success of new music distribution technologies and a growing capacity to provide local content in forms that satisfy the demands of local consumers, in spite of very high levels of unauthorized copying and distribution of physical media, suggest that copyright has not been the key impediment to the success of international artists and labels in the Chinese market.

Formal accession to international copyright treaties, combined with low levels of enforcement and market protection in the form of censorship and media investment regulations, has ensured that China has secured many of the benefits of international trade, without exposing its domestic creative and cultural industries to the rigours of fully fledged international competition. Rather than falling victim to globally dominant exporters of intellectual property, China's domestic music industry is successfully developing a market for local content, alongside its own capacity to provide the content and services demanded by Chinese consumers. This approach fits comfortably with Ha-Joon Chang's (2008) observations about the potential for government intervention to ensure the success of infant industries in developing economies.

However, restrictive cultural policies are not entirely responsible for the failure of major international record labels to capture a major share of the market in China. Unwillingness to adapt to an environment where ownership of intellectual property rights cannot guarantee control over how music is used and distributed, and reluctance to explore alternative approaches to music distribution and licensing have also contributed to their lack of success. There are signs that China's domestic music industry has been able to turn high rates of unauthorized copying and distribution to its advantage, pursuing business strategies capable of succeeding in the context of a weak copyright system and gradually adopting rights licensing systems as the industry becomes better organized, for example in the licensing of rights to the mobile distribution of music and in the apportionment of rights associated with the *Supergirl* series.

The importance of access to a means of controlling the distribution of content and organizing the collection of revenue on a mass scale in spurring large-scale commercial activity within the music industry seems equally obvious. The appearance of mobile music distribution networks in China has been associated with a rapid increase in levels of commercial investment and organization. The apparent association between this distribution bottleneck and the growth of an organized, profitable commercial music industry in China highlights the continuing role of monopoly structures in the commercialization of culture in a digital age. It is possible that the distribution of music through mobile networks will allow just one or two mobile operators to control access to one of the market's few controllable distribution and revenue collection channels – an even higher level of concentration than the handful of major international labels that have dominated the global recorded music industry.

It is tempting to believe that the Chinese government's concern with controlling culture is fading and that new technologies are resulting in greater freedom for producers and consumers of cultural and entertainment products. However, the potential for the Chinese government to exert its influence in shaping the cultural fabric of an emerging, highly centralized commercial music industry should not be forgotten. Baranovitch (2003) has argued that limiting access to economic opportunities for musicians played an important role in the government's control of culture and entertainment in an increasingly commercial environment throughout the 1990s. The continued existence of the publishing licence system and the advent of a single network for the most lucrative form of mass distribution of music in China, in which the government remains the dominant stakeholder, suggest very strongly that the Chinese government remains interested in the power associated with popular culture and is developing methods of exerting its influence in the context of a market-driven economy.

NOTES

1. The IFPI 2006 *Piracy Report* claimed that rates of 'physical piracy' in China were 85 per cent. The total value of the legitimate physical music market was calculated at US$120 million, while the value of physical piracy was reported to be US$410 million (Kennedy 2006). However, when music industry professionals interviewed by the author in Beijing in 2005 were asked to estimate levels of unauthorized copying and distribution of physical music products, estimates were as high as 95 per cent.
2. In October 2001 the PRC government amended China's copyright law in preparation for accession to the World Trade Organization. The copyright amendments attempted to bring the law into line with the WTO's Agreement on Trade-Related Aspects of Intellectual Property Rights. The amended law provides copyright owners with recourse against infringement through administrative channels, court proceedings or a combination of these options. A more detailed discussion of the role of the amended copyright law in the PRC's film and music industries can be found in Montgomery and Fitzgerald (2006).

5. Fashion and consumer entrepreneurs

'I think China's government has realized that the creative industries are very important for China. China was famous for its cheap labour. But right now several other countries are cheaper than China – like Vietnam and India. What should China do to compete in the future? The answer is that it needs to be more creative.'
Zhu Baixi, creative entrepreneur

'You have a much better life if you wear impressive clothes.'
Vivienne Westwood

The past 30 years have seen China transformed from a land of 'blue ants' to a shopper's paradise. Gleaming malls, fashion boutiques, crowded markets and the internet provide urban consumers with access to a dazzling range of fashionable clothing and accessories. For those hungry for guidance on how to navigate this increasingly complex consumer landscape well, and eager to make the 'right' choices about what to wear and how to look good, China's fashion media is also developing quickly. News-stands brim with international titles such as *Vogue* and *Bazaar*, jostling to convince both readers and advertisers of their ability to inform, educate and advise the fashion-conscious about the latest trends, and to explain how astute shoppers might incorporate these trends into their own identity and lifestyle.

China's emergence as the new 'workshop of the world' has had a profound impact on the economies of developed nations. The expansion of China's export manufacturing sector has played a key role in providing the world's consumers with access to inexpensive consumer products, including clothing, at low prices (Kynge 2007). New technologies such as bar-coding and electronic stock-tracking systems have made it possible for retailers to gain faster access to information about what people are buying and what they are leaving on the shelves (Fisher *et al.* 1994: 86). The application of this information in the context of low-cost labour, a ready supply of raw materials and increasingly flexible manufacturing systems, such as those provided by China's textile and garment sectors, has helped to speed up modern fashion cycles and has dramatically increased the variety of products available to consumers around the world.

But, as Ha-Joon Chang (2008) has pointed out, the benefits of competing in a global marketplace on the basis of low-cost manufacturing and cheap labour are limited (Chang 2008). Furthermore, production costs in China are increasing, as the value of the Chinese currency rises (Kwan 2008) and improvements are made in the conditions and pay of factory workers (Au 2009). In 2008 over one-third of China's textile and apparel manufacturers, largely located in the south, shut down as a result of a sharp drop in export demand associated with the 2008–09 global financial crisis. The resulting job losses, which tallied in the hundreds of thousands, and a very real threat of social and political unrest, highlighted the size of the challenge faced by policy-makers as they attempted to manage the transition away from labour-intensive manufacturing towards higher-value-added areas of production, such as design and brand development (Shen Lu 2009).

While markets for cheaply produced Chinese exports are growing less secure, the domestic fashion market is expanding. As a system of signs and symbols associated with status and identity, the language of the global fashion industry is being incorporated into the lives of Chinese citizens, particularly in urban areas. Economic and social transformations are being accompanied by rapid growth in demand for the symbols and experiences of aspiration, affluence and modernity that have become the currency of modern consumer culture across the globe (Farrell *et al.* 2006), and China is on track to become the world's largest consumer of luxury products by 2014 (HKTDC 2008).

Drawing on interviews with Chinese fashion industry professionals conducted in early 2009, this chapter examines the rapid development of an organized fashion industry in China since the 1980s. The global fashion industry relies on copying, peer-to-peer communication through social network markets (Potts *et al.* 2008) and consumer entrepreneurship (Hartley and Montgomery 2009). As Antonia Finnane's (2007) detailed historical study of vestimentary change in China over the past century makes clear, all of these factors played an important role in the clothing choices made by Chinese citizens prior to the economic reforms of the 1980s. Employing Alexei Yurchak's (2002) work on 'Entrepreneurial Governmentality', I argue that fashion values and entrepreneurial modes of thinking in relation to clothing established prior to reform have played a key role in the swift expansion of consumer demand for fashionable clothing and China's rapid incorporation into the global fashion system in the post-reform period.

Circulation of images and ideas related to fashion and consumption has been closely associated with the re-emergence of a fashion industry in China since the early 1980s (Hooper 1998: 169; Finnane 2007: 265). The entire fashion system is based on individual choices within a system where 'choices are determined by the choices of others' – the very definition of a social network market (Hartley 2009: 63). Thus the proliferation of technologies for communicating information about both possibilities and choice to consumers has been at the heart of the development of the global fashion system. Greater opportunities for peer-to-peer communication and consumer access to images and information through the internet might be understood as a continuation of this process, rather than a disruptive technology.

Finally, the chapter examines the role of intellectual property in China's fashion story. Historical studies pointing to the development of fashionable dress in Europe from the fourteenth century[1] provide a clear indication that fashion systems have, in the past, been able to develop in the absence of intellectual property protection. China's experience adds further weight to arguments that intellectual property protection is not essential for the growth of a commercially focused fashion industry (Cox and Jenkins 2005). However, this is not an indication that IP has no role to play in the modern fashion system. The widespread use of trademarks, rather than copyright, by fashion businesses is assisting entrepreneurial consumers eager to capitalize on the value of brand identity.

A POLITICAL PAST

Unlike film and recorded music, which developed as organized areas of cultural production and associated commercial activity during the twentieth century, costume and clothing have served as powerful sites for cultural and political expression and the exercise of state power over the individual throughout China's history. As elsewhere in the world, China has a long and rich tradition of creativity in textile design and the use of clothing and accessories to denote social status and wealth. However, as Beverley Hooper notes, Western scholars have paid little attention to fashion's development in China:

> In few societies has clothing been a more significant symbol of a nation's political culture. Yet apart from the occasional reference by western journalists to China's 'blue ants' in the Mao era and to miniskirts and makeup in more recent years, little use has been made of these artefacts. (Hooper 1994: 164)

Antonia Finnane's (2007) book *Changing Clothes in China* makes a substantial contribution to filling the gap in the literature identified by Hooper in the mid-1990s. Finnane provides the most comprehensive account to date of fashion's role in the everyday lives of Chinese citizens, tracing its roots from the late imperial period, through the early days of the PRC, to the highly varied, globally influenced wardrobes of China's population in the twenty-first century.

In her exploration of vestimentary change in China, Finnane makes a powerful argument that fashion in the post-reform era represents the continuation of patterns of consumption and behaviour evident from at least the late imperial era. While a fashion industry was not yet apparent in late imperial

China, fashion was playing an important role in some segments of Chinese society (Finnane 2007: 2). By the early twentieth century, key features of a modern fashion system, such as rapidly changing dress styles among urban-affluent consumers, were becoming apparent. The early 1920s saw the emergence of a fashion industry, characterized by mechanized textile production, advertisements on billboards and in newspapers, the proliferation of pictorial magazines, fashion designers, retail promotion, competition between local and international products and fashion parades (Finnane 2007: 101).

As with the infant film and recorded music industries of the same period, the development of a fashion industry in China was severely interrupted by civil war and foreign invasion – a period of almost continuous warfare that lasted for 16 years (Selden 1995; Finnane 2007: 206). Following Liberation in 1949, cloth shortages, a national emphasis on the virtue of economy and frugality, and important changes in the status and role of women had a profound impact on China's vestimentary landscape. Rather than resuming the trajectory of internationally influenced fashion development begun prior to the onset of war, China's citizens turned to styles that minimized gender difference, emphasized functionality and provided an outward display of support for the social and ideological goals of New China (Finnane 2007: 201; Wu 2009).

As the title of George Paloczi-Horvath's 1963 political biography of Mao, *Mao Tse Tung: Emperor of the Blue Ants*, suggests, for foreign audiences the subjugation of China's population to communist ideology and parallels with feudal imperialism were encapsulated by images of a vast nation dressed in an apparently uniform style (Paloczi-Horvath 1963: 351–2). Finnane acknowledges that, for outsiders, what people wore was the single most striking feature of communist China (Finnane 2007: 257). However, she challenges assumptions that an apparent lack of variety or obvious change in clothing trends after 1949 resulted from the straightforward subjugation of individuals to the will of an all-powerful leader. Rather, the uniformity in clothing styles during the early years of the PRC was a visible embodiment of complex processes that continued to involve status consciousness, patriotism and the agency of newly liberated women eager to play an equal role in building the nation's future:

> Social status in the early years of the PRC was most recognisable in the form of a male cadre (i.e., party functionary) dressed in a Sun Yatsen suit. Women responded to this vestimentary sign by donning the Lenin suit. Double-breasted with a turn-down collar, modelled on the Russian army uniform, the Lenin suit became popular among women revolutionaries first of all in liberated areas of the north-east, in the late forties. (Finnane 2007: 204)

Images of celebrated popular figures circulated by the media helped to inspire clothing choices, and to create associations between specific styles of clothing and success:

'Aunty' Zheng from Harbin, born around 1940, recalls as a child seeing newspaper photographs of labour heroines such as China's first woman tractor driver, Liang Jun, and first woman truck driver, Tian Guiying: 'They all wore Lenin suits, and looked very smart. I really envied them, and wanted to grow up quickly so that I could wear one, too.'[2] (Finnane 2007: 205)

Ordinary citizens proved surprisingly resistant to centrally orchestrated efforts to influence their clothing choices. In early 1955 the Ministry of Culture authorized a nationwide 'dress reform campaign' aimed at encouraging China's citizens to 'dress up nicely' (*daban piaoliang*) and identifying styles of clothing that might be considered appropriate in the context of New China (Finnane 2007: 206). In spite of nationally coordinated attempts to promote a different dress aesthetic, including nationwide fashion shows in 1956, the campaign made little headway in the face of ideological prejudices and social pressure. The popularity of the cadre suit was simply too deeply connected to the semiotics of status, belonging and aspiration of the times to be dislodged by an official campaign intended to promote greater variety or a more pleasing visual effect.

In any case, proponents of dress reform had little time to realize their ambitions. The acceleration of political developments and the launch of the Great Leap Forward in 1958, a mass campaign intended to leapfrog China into the ranks of the world's industrialized nations in just five years, left little time to consider fashion. The disastrous famine and shortages of resources that followed saw China's population reduced to making their clothes from 'patriotic wool' or, in other words, from scraps (Finnane 2007: 226).

If status consciousness and the choices of individuals about what they *wanted* to wear had played an important role in the early years of New China, extreme ideology and fear of persecution became important factors in sartorial decisions during the Great Proletarian Cultural Revolution (GPCR), which lasted from 1966 to 1976. Red Guards (*hong wei bing*) – a mass movement composed mostly of students – were instructed to 'defeat thoroughly all exploitative old thought, old culture, old customs, and old practices' (Lin Biao, cited in Yan and Gao 1996: 65). The direct link between the ideological position that might be attributed to individuals and the clothing they wore was highlighted by the Red Guards. Eager to demonstrate their revolutionary fervour and passion for realizing Mao's cultural ambitions, the Red Guards subjected individuals who dared to appear in styles of clothing deemed unacceptable, such as narrowly cut trousers, fashionable shoes or a *qipao*, to public humiliation and, often, violence (Finnane 2007: 228; Wu 2009: 2).

In August 1966 the Red Guards of the No. 2 Middle School in Beijing became the first to put up 'big character posters' proclaiming 'A Declaration of War on the Old World'. Their declaration specifically targeted features of dress and appearance associated with the 'young buds of capitalism':

Odd hair styles such as 'aeroplane' and 'spiralling pagoda' and Hong Kong-style jeans and T-shirts, as well as pornographic pictures and publications, must be severely suppressed. We must think that such are small matters, yet the restoration of capitalism begins precisely in these small things. (In Yan and Gao 1996: 66–7)

Rather than marking a radical departure from earlier processes of meaning construction and the communication of individual identity and ambition, during the Cultural Revolution these aspects of clothing became even more important. The narratives that might be associated with a particular cut or style mattered deeply, and clothing choices became a matter, in some cases literally, of life and death (Finnane 2007: 245). According to Wu (2009), who examines fashion in China between 1949 and the present day, during the Cultural Revolution:

dress was a highly visible and tangible target for political attacks, and dressing often became a high-stakes guessing game. But at the same time, dress was also a power-ful tool for protection. In this sense, in the years of the Cultural Revolution, every-one had to be fashion conscious in order to adjust to the nuances and subtle changes in dress and appearance in order to wear the 'correct' fashions. (Wu 2009: 8)

Although scarce access to the resources needed to make new clothes meant that fashion cycles slowed considerably, styles continued to evolve – from the army uniforms which were the height of fashion during the early years of the Cultural Revolution, to the Mao suits that had taken their place by the 1970s. The vestimentary landscape of the Cultural Revolution may have looked monot-onous to outsiders, but from an insider perspective it remained deeply embed-ded in the complex aesthetic, political, social and economic realities. The visual and stylistic flattening of fashion under Mao was accompanied by a deepening of the emotional impact and political implications of sartorial details, and the position of a Mao button or the way a scarf was worn became fashion choices with clear, and sometimes dangerous, implications (Wu 2009: 2).

The persistence of a multifaceted semiotic landscape related to dress, and a level of individual agency in relation to clothing choices, goes some way to explaining the failure of Jiang Qing's attempts to convince the masses to adopt a new style of dress in 1974. In 1974 Mao's wife, Jiang Qing, designed and commissioned what was intended to be a distinctive national dress for Chinese women. In spite of concerted efforts to popularize the new style, and the repeated discounting of the more than 135,000 dresses manufactured at Jiang Qing's request (Finnane 2007: 254), without meaning and context the new style was a failure:

in the hypersensitive political environment of the time, trying out new styles was always risky, no matter how powerful or politically correct the creator of the style might seem. And in this case, a costly, so-called feminine dress that drew inspiration from both the West and China's feudal past made Jiang Qing's creation doubly suspect, impractical and unpopular. (Wu 2009: 5)

By drawing the reader's attention to what clothing meant to individuals, particularly young people eager to impress their peers, Finnane makes an argument for the persistence of fashion values even during a period widely regarded by foreign observers as one in which fashion was entirely displaced by ideology.

Finnane's rereading of China's vestimentary history raises important questions about how 'fashion' is defined. Christopher Breward (2003) observes that, over the past 15 years, fashion has received a growing level of attention from the academic community. But, while the body of critical literature related to fashion is growing, its scope, so far, has been limited:

> These works have tended to concentrate on the creation of fashionable western dress between the fourteenth and nineteenth centuries, seeing in the study of clothing a useful framework for analysing trends in the manufacture, distribution, and use of commodities in historical communities. (Breward 2003: 21)

As Finnane discusses in her opening chapter, many influential theorists and historians of fashion have defined fashion as peculiar to Western societies (Finnane 2007: 6). Failure to find Chinese counterparts to 'fashion' as it has been understood in a European context may simply reflect the failure of English-speaking scholars and historians to pay closer attention to developments in China. However, as a highly complex area of cultural production and consumption associated with status and constant change, in which the consumer choice is itself productive, fashion is frustratingly difficult to pin down. The work of fashion designers and developments in the technologies of manufacturing and retailing undoubtedly play an important role in making fashionable products available to consumers. But, as Breward observes, even when all of these complex factors are taken into account: 'the emergence of a new fashionable style is revealed as a phenomenon which slips tantalisingly between categories and moments in a never-ceasing play between the processes of production and consumption' (Breward 2003: 21).

Important parallels exist between Finnane's exploration of the role of fashion in the lives of Chinese citizens over the past century and Alexei Yurchak's (2002) search for the origins of business practices in post-socialist Russia. Yurchak observes:

> Based on a classical understanding of entrepreneurship, young Soviets in the late 1980s were not supposed to be good at inventing and running private businesses because their generation was raised in a society in which private business was

practically non-existent. And yet, in the late 1980s great numbers of young people quickly started creating new private businesses and turned out to be exceptionally good at it. (Yurchak 2002: 278)

Building on Foucault's concept of governmentality, Yurchak suggests that the skills being used by young Russians in the late 1980s can be attributed to the development of what Yurchak calls 'entrepreneurial governmentality' developed prior to the period of reform. By adopting the analytical framework of entrepreneurial governmentality, Yurchak is able to discuss entrepreneurship and entrepreneurial behaviour in the absence of market-based private businesses.

While the conditions required for the growth of an industrialized fashion industry in China have been absent for much of the past century, Finnane makes a powerful case for the impact of social context, aspiration and constant change on a complex semiotic landscape associated with clothing. Popular resistance to centrally coordinated efforts to reform dress styles under Mao adds further weight to Finnane's argument for the existence of a fashion landscape deeply connected to social network markets and the identities of individuals. As the next sections discuss, these factors help to explain the speed with which commercial fashion activities have appeared, and the strategies of governance that have related to them, in the post-Mao era.

TRANSFORMATION AND A GLOBAL AESTHETIC

Since the very earliest days of reform and opening up, fashion has enjoyed a level of freedom absent in many other areas of cultural production. After Mao's death in 1979, Deng Xiaoping began a long process of market reform and the opening of China to greater interaction with the international community, in the interests of economic reinvigoration. In 1980, just a year after Mao's death, the Central Academy of Art and Design established China's first university course in fashion (Li 2009). Associate professor Li Wei, who began studying fashion at the Central Academy of Art and Design in 1982, describes an environment in which exposure to international influence has been consistently encouraged:

In the 1980s one of the first international designers to come to China was Pierre Cardin. He brought clothing, but he also brought French cuisine, opening the restaurant Maxim's in Beijing. In 1985 Yves Saint Laurent came and did a fashion show in Beijing, and I was lucky enough to be able to work on his show. This was the first time that there had been no distance between a great designer's clothing and us. We were able to see and learn many things, and it opened our minds. Afterwards a great Japanese designer also came and showed in China. . . . Since the 1980s, our school has also frequently asked Japanese, American and French teachers to come and lecture. These people have brought with them international information, ideas and innovation. (Li 2009)

In order to translate international influence from a luxury enjoyed by an elite few within universities into the reinvigoration of the fashion system more widely, a medium for the mass communication of images and ideas was necessary: a fashion media. The 'fashion' magazines that first appeared in China following the end of the Cultural Revolution were not consumer- or market-oriented. They were produced by textile and garment manufacturing or trade companies, higher education institutions and research groups related to the arts of clothing production. According to Finnane (2007: 265), these magazines were intended for fashion and design professionals who required information on manufacturing techniques, export requirements and local and international trends.

The revival of fashion publishing was associated with official efforts to revive China's garment industry and, in the words of party general secretary Hu Yaobang, to 'get Chinese people to wear clothes that are a bit neater, cleaner and better looking' (Finnane 2007: 265). The government lent its support to the establishment of new university degrees and departments, of which the Central Academy of Art and Design's fashion degree was the first (Li 2009), as well as factories, publications and even modelling troupes, as it worked towards building an environment conducive to fashion (Finnane 2007: 265).

By the mid-1980s, regional trends were being disseminated through popular media and cultural products, in particular films, music and television. The first post-reform joint publication of a fashion magazine occurred in 1985, when Japan's Kamakura Bookshop co-published a Chinese edition of its fashion magazine with the China Fashion Magazine Company, featuring patterns, designs and photographs by Japanese contributors. In 1988, following a diplomatic agreement with the French government, a Chinese edition of *Elle* made its debut. Chinese fashion magazines began to shift away from coarse paper with colour inserts and clothing patterns that readers could make at home, towards glossy paper, more visual images, less text and greater focus on the designs of big European fashion houses and international brands (Finnane 2007: 265–6). The International Federation of the Periodical Press (FIPP) estimates that in 2004 around 9000 magazine titles were published in China. In spite of this impressive number, just a handful of foreign titles were dominating the market. The gross advertising revenue of *Cosmopolitan*, *Elle*, *Rui-li* and *Esquire* was RMB1.2 billion, 25 per cent of the total advertising market in China (when measured at rate card) (International Federation of the Periodical Press 2006).

The economic opportunities associated with the global fashion industry have been difficult to ignore: in 2008, the accumulated export revenues of the Chinese apparel industry were US$119.790 billion (China Research Intelligence 2009). In 2006, China became the world's largest apparel exporter (McCormack 2006), and its share of global exports in apparel increased from around 10 per cent in 1990 to around 30 per cent in 2007 (Au 2009). China's total consumption of luxury goods such as jewellery, fashion clothing, leather products and perfume topped US$8 billion in 2007, accounting for 18 per cent of global spending in this area, prompting Yu Guangzhou, vice minister of commerce, to predict that China is likely to become the world's biggest luxury market by 2014 (China Retail News 2008). The growth of domestic consumption is being viewed as central not only to China's continued economic development but also to the stability of the global economy (World Economic Forum 2008).

By 2006, urban-affluent consumers commanded RMB500 billion, nearly 10 per cent of urban disposable income, in spite of accounting for just 1 per cent of the population (Farrell *et al.* 2006: 62). According to global management and consultancy firm McKinsey and Co., these individuals:

> consume globally branded luxury goods voraciously, allowing many companies to succeed in China without significantly modifying their product offerings or the business systems behind them. And since this segment is currently concentrated in the biggest cities, it's easy to serve, both for companies now entering the Chinese market and for old hands seeking a steady revenue stream. (Farrell *et al.* 2006: 62)

Rather than being limited by the peer-to-peer circulation of its core products – ideas, designs and symbols – fashion depends on connection and dispersed distribution to stimulate consumption and to provide the resources and incentives for continued innovation. Visual resources are a vital component of a modern fashion industry. Not only do images provide consumers with models of fashion, but they provide fashion producers with opportunities to showcase their latest offerings. The development of a fashion media, as a medium for the mass circulation of the images and ideas necessary for the construction of a common visual language, was a vital enabling technology in the development of the fashion industries in both the United States and Europe (Hill 2004).

Magazines often act as a theatre for fashion labels in which advertisers, in particular, are able to convey carefully constructed images of their brand. Designers, fashion buyers and fashion stylists depend on the fashion media as a source of inspiration and information about what others are making and the images they are creating. As Wilson observes, for consumers of fashion, words and images not only perpetuate the development of new styles but also assist in the association of products with images of desire and aspiration: 'Since the late nineteenth century, word and image have increasingly propagated style. Images of desire are constantly in circulation; increasingly it has been the image as well as the artefact that the individual has purchased' (E. Wilson, cited in Breward 2000: 29).

In a sophisticated, commercially driven fashion landscape, fashion reflects the aspirations, desires and sensibilities of modern consumers. The choice between different designers or labels allows individuals to identify with the values claimed by brands, capitalizing on images created by designers and labels in the construction of their own identity:

> The Modernists, for example, (represented by, say, Jil Sander) are an entirely differ-
> ent breed than the Sex Machines (Tom Ford for Gucci). The Rebels (Alexander
> McQueen) can easily be distinguished from the Romantics (John Galliano). This is
> not a question of socio-economic status or age. Members of the Status-symbol tribe
> (Marc Jacobs for Louis Vuitton) have neither more nor less money than members
> of the Artistic Avant-Garde (Rei Kawakubo for Commes des Garcons), but they do
> have very different values and lifestyles. (Steele 2000: 7)

Thus, the fashion media plays a vital role in providing makers and consumers with images and concepts that are part of the common visual language of a global fashion landscape. As a result, rather than disrupting this system, new technologies for copying and communication facilitate processes of value creation and innovation.

IP AND THE ENTREPRENEURIAL CONSUMER

Very high rates of unauthorized copying and distribution of copyrighted material have been cited as important factors in the failure of international film and music producers to succeed in the Chinese market. Levels of trade-mark violation are also very high, and consumers have ready access to coun-terfeit versions of famous brands. Some estimates suggest that China accounts for up to 80 per cent of the counterfeit goods found in the global marketplace (Chow 2006: 7). But, while China has been the subject of ongo-ing criticism for its failure to enforce intellectual property law, the market within China for genuine versions of fashionable products continues to increase.

As Jessica Litman observes, the success of the fashion system challenges many common arguments about the role of intellectual property in the creative industries. Litman somewhat facetiously illustrates her point:

> Imagine for a moment that some upstart revolutionary proposed that we eliminate
> all intellectual property protection for fashion design. No longer could a designer

secure federal copyright protection for the cut of a dress or the sleeve of a blouse. Unscrupulous mass-marketers could run off thousands of knock-off copies of any designer's evening ensemble, and flood the marketplace with cheap imitations of haute couture. . . . The dynamic American fashion industry would wither, and its most talented designers would forsake clothing design for some more remunerative calling like litigation. All of us would be forced either to wear last year's garments year in and year out, or to import our clothing from abroad. (Litman 2001, cited in Cox and Jenkins 2005)

As Litman concludes, in fact the fashion industry operates perfectly well in the United States without copyright protection. Fashion is not only able to survive without copyright, but the complex, open-system dynamics on which fashion's cycles of innovation and obsolescence are based could not function without very high levels of freedom to copy and the free circulation of the images and ideas upon which both copies and innovations are based.

The aspect of intellectual property that has been taken up most successfully by firms eager to capitalize on efficiencies of scale within the fashion industry has been trademarks. Copyright and design registration both deal with intellectual property in a narrow form by protecting an original expression fixed in material form. Trademarks, on the other hand, are an area of intellectual property concerned with protecting investments in image and reputation:

Trademarks are symbols used to identify the origin of a product in a commercial context. By identifying the source of the product, trademarks serve an important consumer protection function. In contrast to copyrights and design patents, which are used to protect the artistic and ornamental aspects of a product, trademarks protect only the link between the product and its source, not the product itself. If the makers of knockoff goods affix their own trademarks to their products, then trademarks actually can serve to distinguish knockoff goods from originals and minimize consumer confusion. (Cox and Jenkins 2005)

Copyright places an emphasis on a copyright owner's right to control the distribution of a work and the terms on which it can be used. In contrast, the success of trademarks in the fashion industry stems from their ability to incorporate the entrepreneurial habits of consumers, and their desire to remix the symbols of fashion in ways that will convey a message and identity that they have chosen for themselves.

As mentioned earlier, Yurchak argues that entrepreneurs are individuals who possess an 'entrepreneurial governmentality', a disposition that allows them to understand economic and social relationships in terms of symbolic commodities such as risks, capital, profits, costs, needs and demands (Yurchak 2002). In many ways, fashion 'consumers' behave in exactly this way, seeking out maximum social and semiotic value in return for their investment of time, creativity and money in each of their fashion choices. Trademarks leverage the semiotic value of fashionable products, indexing meaning, signalling to others their price and alluding to the constructed identity of a brand. Thus they provide a tool that entrepreneurial consumers are able to use for their own purposes in a competitive market of status differentiation.

There can be no doubt that China's fashion industry is far from mature. It is developing rapidly, but consumers, designers, editors and photographers have had less than 30 years to become literate in the highly complex global system that relates to the international fashion industry. Although creativity is rewarded in fashion, there is a fine line between success and failure in fashion innovation. Design literacy and a highly developed understanding of the language of fashion are valuable skills for those attempting to navigate a tricky landscape of taste. In 2005, Hung Huang, CEO of China Interactive Media Group, expressed her frustration at the difficulty of finding stylists who understood the fine line between innovation and failure in fashion:

> We're trying desperately to look for someone who can be edgy and beautiful and artistic and without immediately going into trash. We love to hire people who majored in art history. . . . Photographers are not the problem; they are technically skilled. The problem is that . . . the stylists can't figure out the difference between sexy and obscene. They struggle to negotiate the lines between sexy and obscene, between beautiful and plastic, because they don't have details. (Hung 2005)

If China's fashion industry professionals are struggling to master the highly specific language of beauty, taste and style in the context of the twenty-first century's global fashion system, it is not surprising that consumers also find this task difficult. As a result, the power of brands to simplify the language of fashion and its associations with status is highly valued. According to Chen Xiang, co-founder of the innovative Chinese brand Decoster:

> Every city has one premiere shopping mall. They hope that on the first floor they will be able to have Louis Vuitton and Gucci – these kinds of big luxury brands. Not all Chinese consumers are able to go shopping in Hong Kong, Europe or Japan. Some extreme consumers in China don't care about the price if they like certain brands. As a result, in the Chinese market, Louis Vuitton is more popular than Zara. The number of Louis Vuitton shops is higher than the number of shops selling Zara. I think this is part of a process of maturation for Chinese consumers. On the other hand, creative brands are unable to open boutiques in the way that Louis Vuitton does, because most Chinese consumers are unable to mix and match their own clothes. (Chen 2009)

As Chen Xiang implies, consumer confidence in mixing and remixing elements of fashion made available to them from a wide variety of fashion producers, and their understanding of the value of their own creativity and skill in re-creating a look or supplementing their identity, is heavily dependent on the sophistication of their understanding of the symbolic subtleties of the modern fashion landscape. As China's consumers continue to be exposed to global fashion, it can be expected that they will become more discriminating in their consumption of fashion products, and that the market for well-designed but less exclusive clothes will expand.

There are signs that the younger generation, who have grown up in a much richer and better-connected consumer environment than their parents, are already doing this. Nonetheless, it is their parents who command the greatest purchasing power. Traditional forms of fashion media, such as fashion magazines, continue to play a highly influential role in the development of China's fashion landscape:

> Young people get information through the internet, for example they buy Japanese street brands online. But the groups who have real purchasing power get information from traditional media: TV, magazines and newspapers. New magazines offer free ads to Louis Vuitton. (Chen 2009)

Even among more conservative consumers, sophistication in perceptions of brand value is growing:

> If my friend has a big-brand bag, I might want to have the same one. But right now, this situation is already changing. If you have a big-brand bag, I want a different big-brand bag. When the children of the current big-spending consumers grow up, they won't need brands like Louis Vuitton. (Chen 2009)

Even if the next generation of Chinese consumers is less enamoured with Louis Vuitton, it seems likely that other brands will still have a role to play. The manufacturing and consumption of physical artefacts remains at the heart of the fashion industry. Digital technologies have challenged distribution monopolies in the film and music industries because they have made it possible to make seemingly infinite numbers of perfect copies of film and music products at almost no cost. Although these technologies have increased access to information about fashion and the choices of others, and provided additional opportunities for the purchase of fashionable products through online shopping, they have not dramatically altered the costs of physical production or the need for skilled labour, raw materials and factories capable of making often highly sophisticated physical objects.

While counterfeiting arguably allows for some manufacturers and retailers to profit from a creative process that they have not invested in, it has also been argued that the impact of this form of copying is low because counterfeited apparel and accessories are largely purchased by sections of the market who would be unlikely to purchase the genuine article (Barnett 2005). Regardless of how cheap labour is or how quickly transport costs decline, turning any design into a physical product that can be sold and worn involves costs. High-end producers are able to employ expensive materials and labour-intensive detailing in order to differentiate their products from less expensive copies. Globally recognized luxury brands also invest in creating shopping 'experiences' for consumers that reinforce the exclusivity and luxury values that their brands are associated with (Okonkwo 2007: 86).

Furthermore, there are signs that Chinese consumers value the information and reassurance that branding makes available to them. In January 2009, the *Mirror* reported on plans to open a shopping mall dedicated to 'fake' brands in Shanghai, such as McDoanald's, a Starbucks-style coffee shop Bucksstar Coffee, and Pizza Huh. An 'angry shopper' was quoted as saying: 'Not every shopper is brand conscious so a lot of people will walk into these stores thinking they are getting the real thing' (Wood 2009). This incident highlights the key difference between the deliberate purchase of a counterfeit product, which informed consumers may hope will fool their friends, and the vulnerability of consumers who believe that they are paying for an 'international' brand or experience which is, in fact, no more than a local impostor.

However, in developed fashion markets, 'fashion' and 'luxury products' are not interchangeable concepts. As Chen Xiang observed, luxury brands are currently over-represented in China's fashion market. According to the Hong Kong Trade Development Commission, while 80 per cent of the world's top-brand luxury products are available in China, Chinese consumers have not yet developed a comprehensive understanding of luxury culture:

> The nouveau riche class lacks knowledge about the brand culture of luxury products, consuming to show off their wealth; and some young white-collar people need an overdraft in order to realize their 'dream of luxury consumption'. Only when buying luxury goods becomes a life style pursued by the Chinese but not a way to display their wealth, can China have a mature luxury consumption market. (HKTDC 2008)

Nonetheless, the emergence of a 'creative class' in China is being accompanied by greater demand for affordable, well-designed fashionable products. Fougere 2, located in Shanghai's Red Town creative cluster, was established in 2008 by a consortium of six young fashion industry professionals. This fashion start-up aims to offer fashionable clothing to Shanghai's creative workforce: models, photographers, graphic artists and stylists (Zhu Baixi 2009). Fougere 2's target consumers are conscious of the value of creativity in fashion, but unlikely to enjoy a salary sufficient to fund the purchase of imported designer wardrobes. Their level of fashion literacy is such that they would prefer to invest in a less expensive original product from a local designer with creative credentials than in counterfeit versions of luxury designer labels (Zhu 2009). Fougere 2 provides China's young creative class with an impressive website that incorporates stylish photography, an online community and a fashion blog: http://www.fj-2.com/. The site allows clients to select an item from a range of original designs and, if they wish, to request that it be customized or altered to suit their preferences.

Local brands are also successfully targeting wealthier Chinese consumers. Decoster emphasizes creativity and originality. In the ten years since it was founded by a husband-and-wife team, Decoster has made the transition from marketing itself to white-collar workers, to targeting a wealthier clientele:

> our customers are those who have purchasing power, who own companies or work as freelancers or have money. The philosophy for a brand is very important. It should have uniqueness and difference, which will give its product high value and help to secure profit. But there shouldn't be too great a gap between a brand and its customers. Otherwise, it is not possible to sell products. (Chen 2009)

While Decoster's style is an international one, it makes no effort to conceal its local origins and deliberately includes Chinese design elements in each season's collection (Chen 2009). Having successfully opened stores in all of China's first-tier cities, such as Beijing and Shanghai, and opening a store in Japan, Decoster is now expanding into China's second-tier cities. As co-founder Chen Xiang observes:

> Not all foreign brands are successful in China. It's a process of development, from welcoming foreign brands, towards Chinese customers developing an understanding that not all foreign brands are good. Retail channels like department stores and shopping malls have been very eager to welcome foreign brands. Stores offer foreign brands good deals on rent and provide them with the best locations within shopping complexes. But, after several years, retail outlets have begun to understand that not all foreign brands are successful. In the past two years, department stores have started to pay more attention to domestic brands. (Chen 2009)

Local governments in China are also making efforts to support the development of the fashion industry. The need to shift away from low-cost manufacturing, towards higher-value-added areas of creative production has been made more urgent by the 2008/09 global financial crisis:

> Different cities have different policies. Shenzhen and Hangzhou are better than Shanghai. Since the most recent economic crisis began, China's government has been paying more attention to the creative industries. Since last year, we have felt that the government has been paying close attention to us. . . . Last year Beijing's fashion association told me that, if Chinese brands want to join foreign fashion weeks, the government will provide sponsorship money. The government will provide RMB2 million to support domestic brands in showing at foreign fashion weeks. The Shanghai government also offers tax-free status for new fashion studios. (Chen 2009)

The impact of efforts to realize China's potential in high-value areas of the fashion industry is also being felt in fashion education:

> The Chinese government is paying more attention to fashion education. For example, the Beijing Institute of Fashion Technology is a local government project. Every local government is setting up a locally branded school of fashion design. The government is paying more attention because apparel is a big market. They believe that more education, more companies and more talent will improve the local economy. (Li 2009)

CONCLUSION

Important elements of continuity exist between the role of clothing as a site of social, political and cultural meaning and continuous change in China over the past hundred years, and the rapid growth in demand for the products of a highly international commercial fashion industry that has become apparent since the early 1980s. The processes of economic reform initiated by Deng Xiaoping have provided a new context for the expression of status-based dynamics and entrepreneurial modes of thinking in relation to clothing. Whether by accident or design, ideal conditions for the growth of a fashion industry are present in China. A rapidly expanding fashion media, rising disposable incomes and increased consumer access to global images of consumption, as well as manufacturing capacity and the state's desire to secure China's status as an influential member of the global community, are all contributing to a rapid increase in demand for fashionable products.

Greater access to the images and products of a globally oriented fashion landscape are widening the semiotic vocabulary of China's fashion consumers and the industry that is developing around them. Although it is taking time to develop local skill in areas such as design, photography and publishing, demand for fashion courses at universities has grown rapidly since their establishment in the early 1980s (Jia 2009; Li 2009; Liu 2009). There seems to be every reason to believe that, as China continues to develop, its capacity to capitalize on growing domestic demand in the fashion sector will also increase.

Modernization, urbanization and the growth of consumer culture are being accompanied by the emergence of a dedicated fashion media capable of communicating fashion values and brand ideals to a widening social base of increasingly affluent and self-directed Chinese consumers. The growth of this area of media production is being given greater traction by the skill and knowledge transfer associated with foreign–local co-publications of fashion titles. Local publishers now have the freedom to team up with experienced international publishing houses to co-publish Chinese editions of titles such as *Vogue,*

Harper's Bazaar and *Rui-li*. In 2007, 59 foreign consumer periodicals published versions in cooperation with Chinese partners, including 26 international fashion and lifestyle magazines (Wang 2008).

The continued expansion of the fashion industry globally challenges many widely posited arguments about the critical role of intellectual property protection for continued innovation and profitability in the creative industries. Market segmentation and the fashion system's ability to reward consumers for creativity and innovation in the context of trend-driven cycles of demand and obsolescence are allowing fashion to flourish in China, where the regulatory environment is challenging traditional film and music business models, which depend on IP to limit copying and control distribution.

What appears to be most important for continued processes of creativity, innovation and growth in consumer demand in relation to fashion is not control of distribution channels or an ability to dictate terms of use, but access to information. The circulation of information makes scalability possible in the fashion industry, by ensuring the dissemination of a common visual language that can then be appropriated by both producers and consumers. Copying helps to signal popularity among consumers, but, because the social nature of fashion consumption ensures that novelty and exclusivity are also valued in the fashion system, copying also helps to generate demand for new designs.

In contrast to copyright, trademarks have been employed by fashion businesses to help create symbolic carriers of information about a product and its price, origins and imagined identity, which can be used by consumers in entrepreneurial ways. The symbols provided by the global fashion system can be employed by entrepreneurial consumers to enhance their own social goals, publically projected identities and desire for status and enjoyment. All of these factors suggest that fashion is an area of cultural and economic activity that deserves greater attention as the creative industries struggle to develop business models capable of succeeding in a digital age.

NOTES

1. See Lemire and Iello (2008) and Ewan (1990).
2. 'Fuzhuang: yiqu bu fan de Leiningzhuang' [Clothing: the Lenin suit – once gone, never to return], *Guoji Xianqu Baodao*, 20 March 2006, news.xinhuanet.com/herald/2006/03/20/content_4322357.htm (accessed 1 October 2006), cited by Finnane (2007: 205).

6. Does weaker copyright mean stronger creative industries? Some lessons from China[1]

INTRODUCTION

In the second half of the twentieth century 'core copyright industries' such as film and music have been closely associated with a rhetoric that asserts that high levels of copyright protection are crucial to the existence of and economic contribution made by this sector of the economy (Boyle 2004). Although the creative industries are much larger than the copyright industries alone, core copyright industries make up a significant proportion of the activities that now fall within creative industries policy frameworks. Examples of activities that are considered to be part of both the creative industries and the copyright industries include film, television, music and publishing, as well as computer software and interactive games (Allen Consulting Group 2001). This overlap between the creative industries and the copyright industries means that it is tempting to conclude that arguments put forward for the expansion of copyright in order to promote growth in the copyright industries should also be applied to the creative industries as a whole.

However, the history of copyright law is one of contestation and debate over the extent to which granting a monopoly right to 'authors' produces either economic or social benefits.[2] A growing body of literature on the economics of intellectual property suggests that the expansion of intellectual property rights suppresses innovation and favours the interests of a few players within the creative economy at the expense of the majority. Boldrin and Levine (2002, 2005, 2008), two highly respected economic theorists, argue that there is no economic justification in theory or evidence for the 'intellectual monopoly' created by copyright and patent law, and advocate the complete abolition of these systems. A similar line is taken by van Schijndel and Smiers (2005), who propose a radical re-formation of the copyright system, which they argue is systematically failing the creative producers it is intended to support.

This chapter explores economic aspects of the relationship between intellectual property protection and growth in the creative industries. It focuses, in particular, on the impact of the global nature of the creative industries on the

93

economic effects of intellectual property rights. The realities of operating in a global rather than national environment mean that it is often more efficient for creative industries firms to adopt strategies that are capable of operating in the context of weak intellectual property rights, even if their home market is one in which these rights are strong. Furthermore, there are important difficulties associated with attempting to apply economic models that focus on the value of *new ideas* to the creative industries, where economic and social value arise from the *reuse of creative works*. The chapter concludes that, because overly restrictive intellectual property regimes act to prevent interactions between creative producers and restrict access to creative resources that can be built upon, weak intellectual property rights favour the growth of the creative industries.

ECONOMIC ARGUMENTS FOR AND AGAINST IP

Conventional theoretical approaches to the economics of intellectual property consider the challenge as one of balancing the economic costs and benefits of granting a monopoly interest in a non-rivalrous good on the production and dissemination of new ideas. New ideas are said to be non-rivalrous because they can be used by many people at the same time (Romer 2002). As a result, the cost of reproducing a new idea once it has been discovered is zero. It is argued that new ideas produce social benefits. But because it is less costly to simply copy ideas than to produce innovations, new ideas are undersupplied in competitive markets (Hirshleifer 1971). The failure of a competitive market to produce an adequate supply of new ideas is generally tackled in one of two ways: 1) through public funding for the production of ideas, as in the form of government funding for research and development and support for the arts; or 2) through the creation of artificial imperfect competition in the form of a temporary monopoly to enable the idea producer to recoup, through monopoly rents, the fixed cost of production.

As such, intellectual property rights are understood as a mechanism through which market failure can be addressed and the supply of new ideas increased. In spite of their market distortion effects, they are widely favoured by economists and policymakers because they preserve market incentives for innovation. Dixon and Greenhalgh summarize:

> To give people an incentive to produce socially desirable new innovations, IPR allow the creators of a nonrival good to appropriate the returns of their innovation for themselves alone. But since IPR make a nonrival good excludable, it constitutes an inefficiency, since the price of the good will be above the marginal cost of producing the good. . . . Economists are then left to adjudicate as to the desirability of using IPR as a spur to innovation, and as an instigator of monopolistic inefficiency. (Dixon and Greenhalgh 2002)

The standard example of intellectual property rights creating an incentive for investment in socially beneficial innovation is new drug development, the costs of which may run to hundreds of millions of dollars. In this instance, intellectual property rights provide an incentive for the production of a socially valuable good (new ideas and processes that give society access to more effective medicines). The trade-off is that a market distortion is created while a monopoly on the use of the new ideas exists. This means that the cost of buying a drug that is still protected by patents will be higher than the cost of simply manufacturing the drug once it has been discovered and developed. Patent protection ensures that investors in the expensive process of developing new drugs are able to recoup the costs of innovation and derive profits from their investment. Once patents have expired anyone is allowed to use the new idea, and so the cost of purchasing the drug declines, ensuring that the benefits of pharmacological advances are enjoyed by as many people as possible. In principle, an optimal level of IP protection balances economic costs and social benefits at both micro and aggregate levels (Gilbert and Shapiro 1990; Cohen *et al.* 2000).

However, the economics of intellectual property have come under scrutiny in recent years for a number of reasons. There is growing recognition of the fact that different kinds of creative activity are associated with sometimes extreme differences in both fixed costs and opportunity costs. While economic theory may support mechanisms through which the costs of innovation are compensated, intellectual property law as it currently exists produces excessive market distortions and is an inefficient mechanism for encouraging innovation (Tabarrok 2002). Many economists view copyright (especially within the creative industries) as a pure source of rent and distortion, rather than a solution to the problem of an undersupply of creative innovation (Murphy *et al.* 2002; Romer 2002).[3]

Boldrin and Levine (2002, 2005, 2008), who base their work on Liebowitz (1985), are particularly critical of widely accepted economic arguments for copyright protection. These authors suggest that creators' rights to own and sell their creative work need to be distinguished from *intellectual monopoly*, which relates to the right to control the way in which a creative work is used once it has been sold. As they explain:

> Intellectual property has two components. One is the right to own and sell ideas. The other is the right to control the use of those ideas after sale. The first, sometimes called the right of first sale, we view as essential. The second, which we refer to as downstream licensing, we view as economically dangerous. (Boldrin and Levine 2002: 209)

Although Boldrin and Levine support the right of first sale, which allows creators to profit from their labour, they believe that monopoly rights to control downstream use simply distort the market and stifle innovation. These rights are often justified with rhetorical claims that they will increase innovation and secure the livelihoods of creators. However, Boldrin and Levine believe that rights to downstream licensing are a result of the rent-seeking behaviour of would-be monopolists seeking to profit at the expense of public prosperity (Boldrin and Levine 2008: 244). They conclude that intellectual monopoly is, in most circumstances, a highly inefficient mechanism for achieving rent creation for idea producers and suppresses innovation. They argue that the most effective course of action would be to abolish both the copyright and the patent systems as they currently exist (Boldrin and Levine 2008).

INTELLECTUAL PROPERTY AND THE CREATIVE INDUSTRIES

Any discussion about the role of intellectual property in the creative industries brings to the fore an obvious difficulty, namely that widely accepted definitions of the creative industries refer directly to intellectual property as a mechanism for the conversion of creative activities into economic benefit (DCMS 1998). As Chapter 2 discussed, an alternative and potentially much more useful approach to understanding the economic characteristics of the creative industries may follow from their conceptualization in terms of their social network market functions (Potts *et al.* 2008), which operate as part of an *attention economy*, not a *property economy*. According to this model, the economic value of the creative industries lies in the capacity of this sector of the economy to drive wealth creation in the broader economy through dynamic processes that involve not just the origination of new ideas but their revision and adaptation and the dissemination of new knowledge throughout the economy.

Furthermore, in an era in which content and culture flow globally, copyright faces an almost impossible task. While international trades in cultural and intangible goods now form an increasingly valuable portion of the global economy, the concepts upon which copyright relies, 'creativity' and 'originality', are difficult to transplant into new cultural, economic, artistic, social and technological contexts (Alford 1995; Mertha 2005; Liu 2006). Given the power of new technologies to break down physical barriers previously associated with creation, distribution and use, it is therefore important to reconsider the extent to which copyright is in fact either necessary to or useful for the development of the creative industries.

The creative industries serve as a useful test-bed for theoretical developments relating to the role of intellectual property in a knowledge-based society. This is because the creative industries are representative of sectors that are predominantly *global* and that derive value from the exploitation and *reuse* of ideas as much as from the creation of globally novel content and work. They are also part of the attention economy (Lanham 2006) and as a result display a high level of *business model flexibility*. This might seem a rather narrow set, representing somewhere between one-twentieth and one-tenth of all economic value creation. But it is also a rapidly expanding class of activity, growing at about twice the rate of the aggregate economy (Potts and Cunningham 2008). These three characteristics raise important questions about how relationships between economics and law in a knowledge-based society are understood.

The next sections of this chapter discuss the impact of these characteristics on the way in which intellectual property law has an impact on the creative industries. They consider, in particular: the impact of globalization, and specifically the internet and information and communication technologies, on the ways in which creative production occurs and the creative industries function; the relative economic value of reusing creative content, compared with the incentive to create provided by monopoly rights; and the co-evolution of business models and institutions in response to technological change.

CREATIVE INDUSTRIES AS GLOBAL INDUSTRIES

The creative industries operate in the context of global flows of information, content and ideas. Although there can be little doubt that these are an important source of dynamism, they also have powerful consequences for the ways in which intellectual property protection has an impact on the creative economy. In the creative industries most firms and consumers either produce or consume, or both, in global markets. Yet, in doing so, they are governed by national laws. Although efforts have been made to harmonize intellectual property regimes globally, enforcement depends on nation-based authorities and infrastructure (Liu 2006). While globalization provides opportunities for value creation in production collaboration and specialization, along with the general benefits of large markets, it also means that creative industries businesses are unable to avoid the dissemination of their products within markets with weak intellectual property systems, where enforcing intellectual property rights may be prohibitively complex or expensive.

As a result, firms operating in global markets are often unable to formulate strategies and business models based on uniformly high levels of intellectual property protection. Business models that depend on a firm's ability to enforce its intellectual property rights quickly and cheaply are cost-effective only in markets in which these conditions exist. This means that the global reach of business models that rely on high levels of intellectual property protection is limited, particularly in relation to key emerging markets, such as China. One response to this situation has been an attempt by developed economies with strong intellectual property systems to create global frameworks for the protection of intellectual property rights and to require nations seeking access to international communities of trade to strengthen national intellectual property systems (Maskus 2000; Wang 2003; Miller *et al.* 2005).

However, as China demonstrates, developing an intellectual property system takes time. Although legislation can be created relatively quickly enforcement is a much more complex challenge for both policymakers and copyright owners. China is far from the only nation in which globalization and new technologies are associated with rapidly increasing access to creative content, images, sounds and information, but levels of copyright protection remain very low. In effect, a constant state of disequilibrium exists in the strength of intellectual property law operating in different national contexts. It is therefore economically rational for firms targeting global markets to formulate strategies based on an assumption that levels of intellectual property protection are low in *all* markets. Such strategies may involve an emphasis on an experience rather than the sale of physical products that can be easily copied, for example live music performances, 3D films that are best enjoyed in a cinema, or multiplayer online games played on closed platforms in real time. They might also involve an emphasis of the status and identity associated with consuming products made by a particular firm or in a specific location, as in the purchase of a luxury branded handbag or consumption of French champagne.

This gives rise to a curious economic property of the interaction between intellectual property law, global markets and business strategies, namely that the presence of strong, effective and efficient intellectual property law in individual territories *may not benefit the creative industries*. This is because, for businesses formulating strategies for global markets, strong intellectual property law matters only if it is available globally. If it is not available globally, then firms have little choice but to alter their business strategies in order to take advantage of opportunities in markets where high levels of intellectual property protection are absent. Because a state of constant disequilibrium exists in the levels of intellectual property protection that relate to global markets, effective global strategies must take into account the aggregate global costs of enforcing intellectual property rights.

In spite of the lack of equilibrium in levels of intellectual property protection in the global marketplace, the creative industries are growing at about twice the rate of the aggregate economy (Potts and Cunningham 2008). Recognition of the global nature of the creative industries and the national nature of intellectual property protection helps to explain why business models that have proven successful in the United States and Western Europe, such as those of the major record labels, have made so little headway in China. The failure of these business models is not a result of a causal connection between the growth of the creative industries and levels of intellectual property protection: as this book has discussed, China's creative industries are developing quickly. Rather, it relates specifically to the inability of business models that depend on high levels of intellectual property protection and enforcement to function effectively in truly global markets.

INCENTIVES FOR CREATIVITY AND VALUE FROM REUSE

The creative industries' concern with the origination and trade of creative *works*, which are the subject of copyright protection, rather than *ideas*, which copyright does not protect, highlights an important disconnect between economic theories of innovation and copyright law. The economics of ideas and information and the growth and market theory that this area of economics is based upon are concerned with the economic value of *ideas* rather than works. While economic models understand the production of new ideas as a significant source of value (Dopfer and Potts 2008) and focus on the impact of monopoly rights on their production, copyright law is explicitly concerned with protecting works, not ideas. This suggests that commonly applied economic arguments for copyright protection, which are based on the economics of ideas, are fundamentally flawed.

The economic justifications for intellectual property protection outlined earlier in this chapter focus on intellectual property's capacity to provide incentives for value creation driven by the *origination of new ideas*. While this form of innovation is undoubtedly important, focusing exclusively on the value of new ideas has resulted in failure to consider value creation associated with other processes, such as reuse (Boldrin and Levine 2002). However, economic value in the creative industries is significantly composed of, if not dominated by, value creation in reuse. Thus the creative industries provide an important challenge to economic theories which focus exclusively on economic value associated with origination and which neglect or ignore the value of reuse.

In contrast to the economics of ideas, which is based on equilibrium models of a closed economic system, evolutionary economics approaches the economy as a complex open system. As Dopfer and Potts explain:

> the complexity of the economic system is due to its *modularity, openness* and *hierarchic depth*. The economic system is modular in the sense of being made up of a large number of functionally specific parts. It is open in the sense that these parts

interact with degrees of freedom. And it is deep in the sense that each module is itself a complex system: every part is whole and every whole is part. (Dopfer and Potts 2004: 3)

Evolutionary economists approach wealth creation as an adaptive process in which solutions to economic problems are found through processes of differentiation, selection and amplification that involve physical technologies, social technologies and business models (Beinhocker 2006). The trial of a wide variety of candidate designs in a particular environment, the selection of the best designs for a particular purpose and the spread of these new designs help to drive wealth creation (Beinhocker 2006: 14). Viewed through this lens, novelty creation may not be the most important aspect of the economic problem that copyright law attempts to address. It may not be the origination of new ideas, but rather the *reuse* of a particular instantiation of an idea, that provides the greatest social and economic value in the creative industries.

The explosion in online creative content since the launch of the World Wide Web in 1990 has demonstrated very clearly that creativity is not in short supply. The creative potential of new technologies has been taken up with enthusiasm by internet users, and the vast majority of those who write blogs, upload photographs, share content and participate in online communities do so without any hope of direct financial reward. The ample supply of creative content and new ideas made available through the internet is a result of the shifting opportunity costs of creative behaviour associated with rising incomes and the mass adoption of tools for creative production, rather than incentives for innovation provided by intellectual property law (Towse 2001). This suggests that creativity is not incentive-constrained under perfect competition and that the standard market failure model of creative supply is seriously flawed, at least in relation to the creative industries.

Although it is possible to imagine new inventions that might be brought to the market in a form that never needs to be revised or adapted for new uses or contexts (for example in pharmacology or biotechnology), this kind of knowledge production is rare. In the creative industries, in particular, it is much more common for new ideas to be made available, taken up, revised, applied to new contexts and revised again. The challenge for firms operating in the creative industries is not an undersupply of creativity and new ideas, as economic theories of intellectual property assume, but the challenge of identifying the products, services and business models best suited to the highly connected, global markets of the twenty-first-century knowledge economy. In this context, the diffusion of ideas and their adaptation to suit the specific context in which they might be applied are important factors in value creation. An ability to access, reuse and alter creative works is a vital component of these processes of innovation and knowledge growth.

Intellectual property law deals with the reuse of works by allowing copyright owners to dictate terms of use and to negotiate downstream licences. However, this is a market model of reuse in production as a factor input or reuse in consumption as utility, but not of *reuse in innovation*, or of reuse in which production is by definition novelty as, for example, in content production. This is an important and much overlooked distinction that is highlighted by the creative industries. Although the creative industries certainly produce value by being creative in the classic sense of producing new ideas and thus novelty, a substantial portion of the value that these industries generate results from the reuse of content in new mediums, contexts and formats.

On the face of it, reuse might appear to be little more than a form of replication that, in a dynamic system, leads to standardization as the most popular ideas dominate the market. However, in reality each instance of reuse occurs within a unique context that includes complex networks of other ideas. The net result is that reusing a particular instantiation of an idea in new contexts and in conjunction with new combinations of other works and ideas *increases* variety. And with that comes exploration of entrepreneurial opportunity space, which is simultaneously a private and a public good. This variety-increasing reuse is deeply ingrained in the creative industries: jazz improvisation, the editing and remixing of content associated with YouTube, and a fashion consumer's selection of a 'fashionable' ensemble are just three examples.

Reuse is also connected to the growth of knowledge through the transfer of ideas and information between different industries. This may occur when ideas developed in one domain, for example chemistry, are applied in another domain, such as biology. It is also an essential driver in the development and commercialization of transformative technologies, such as the internet. However, it is not a process that is confined to theoretical knowledge or ideas. It also occurs in relation to creative works, for example when one piece of visual art is reused or recontextualized in the creation of new art. It also occurs when content from one domain is used in another, such as when visual art is reused in advertising, when music is used as ring tones for mobile devices or when cinematic content is reused in the development of the digital economy.

The growth of knowledge through the reuse of ideas across domains is a defining feature of the creative industries. Current technologies and markets help to shape future technologies and markets in the creative industries, as in other areas of the economy. However, in the creative industries developments *across* domains of knowledge and markets are as important as, if not more important than, past paths within domains of knowledge and markets. In the creative industries reuse is associated with a new and, to date, little considered economic and socio-cultural dynamic that allows the value of new ideas, technologies and creative works to be maximized. This is an important mechanism for the realization of the economic value of both cultural production and digital affordance and may, in fact, be the underlying driver of the growth effects of the creative industries (Potts and Cunningham 2008; Potts *et al.* 2008).

The ability to easily build on the creative works of others and to draw on a global pool of content (both instances of reuse) is potentially the most powerful benefit of the internet for creative workers and industries. Open innovation and production models rely on, and indeed exploit, the reuse of knowledge and, in doing so, explore the space of possible business models premised on such assumptions. Some practical reuse solutions to the restrictive nature of copyright in a digital environment have already been found – notably the General Public License (GPL) and Creative Commons (CC). Yet efforts to harmonize the creative and collaborative potential of new technologies with existing copyright frameworks do not solve the broader problems associated with the copyright system itself.

Movements such as Creative Commons provide opportunities for creators to make, share and remix works legally. They offer creators standardized licences and attach metadata to licensed works in order to ensure that the terms upon which a work can be used are easily identified in a digital environment. While Creative Commons licences assist amateur creators in asserting their right to decide the terms on which their work is used, they do little to lessen the separation between creative resources generated by amateurs and commercial potential that might be realized by entrepreneurs and the business community (Moller 2005; Hancock 2006). Thus movements such as Creative Commons are not an adequate substitute for a more radical reformulation of copyright law to allow the economic value of reuse to be more fully realized (Lessig 2004).

What these arguments amount to is that the reuse of ideas is an overwhelmingly dominant source of economic value in the creative economy. While there can, of course, be no reuse of an idea without an idea's initial creation, legal and economic conceptualizations of the value of new ideas need to recognize that economic value is not simply created at the point of origination. Rather, it accrues through an ongoing process of adoption and adaptation (Dopfer and Potts 2008) in which the value of an idea is realized as it is combined with other ideas, placed in new contexts and used in new ways.

THE CO-EVOLUTION OF PHYSICAL TECHNOLOGIES, BUSINESS MODELS AND LEGAL SYSTEMS

The third characteristic of the creative industries that is often overlooked in debates about the economic impact of copyright law is that business models co-evolve with physical technologies and institutional environments. This has important implications for the impact of high levels of intellectual monopoly on the capacity of businesses to succeed in global markets. Paradoxically, the firms most likely to succeed in these markets are those whose business strategies have evolved in the context of weak intellectual property systems. This fact helps to explain the truly global success of the fashion industry, which has developed effective strategies for the commercialization of creative works in the context of high levels of copying among both producers and consumers. It also helps to explain the difficulty that major record labels have experienced in their attempts to capitalize on opportunities in markets in which copyright cannot be easily enforced.

The co-evolution of business models, physical technologies and legal institutions is clearly illustrated when the development of the recorded music industry in China is compared to its development in the United States. In the early days of recorded music in the United States, highly specialized equipment was required to turn sounds into physical products that could be sold in a mass market. Making multiple copies required hardware that was not widely available. As a result it was relatively inexpensive to control and monitor the production and distribution of music products (Gronow and Saunio 1999). The creation of neighbouring rights made it possible for firms to own the copyright in sound recordings they had commissioned (Laing 2002: 185). Developments in physical technology, the existence of intellectual property rights and an ability to enforce these rights efficiently created commercial opportunities for businesses willing to invest in the production and promotion of music that could be sold to a mass market.

The dominant business model in the recorded music industry in the second half of the twentieth century reflected the technological and institutional environment within which businesses had been formed and developed. Record labels provided artists with access to recording equipment, mass production and distribution channels, and marketing and promotion services, and remunerated them on a royalty basis. Artists received (and still do) income from royalties generated each time a copy of a recording was sold or broadcast. Although developments in physical technology, such as cassette tapes and recorders, presented challenges to the industry's ability to control copying, these changes occurred *after* markets, industry structures, professional organizations and group collection infrastructures had become established. Thus the recorded music industry was generally able to respond in a systematic way, and incremental developments in analogue technologies of copying did little to disrupt its overall structure (Frith 2004).

In China, on the other hand, technologies for mass reproduction and consumption of recorded music became available in the absence of copyright law, an organized domestic music industry, or clear legitimate channels for the distribution of most foreign content. These technologies also became available as China was making the transition from a planned economy to a market system. High levels of demand for popular music, combined with readily available technologies for mass reproduction and consumption and an absence of legitimate distribution channels, contributed significantly to the rise of a black market in music products and highly sophisticated illegal distribution networks (de Kloet 2002). The internet, personal computers and cheap MP3 players have compounded the difficulties associated with controlling distribution: technologies that are challenging approaches to the control and monetization of content globally.

Almost all of the music downloaded from the internet onto personal computers or portable devices such as MP3 players in China occurs without permission from or payment to copyright owners (Daniel 2007; Music 2.0 2008). Not only are new technologies being adopted with enormous speed across China, but they are being embraced fastest by groups traditionally considered most likely to pay for music. Young, educated city-dwellers with relatively high disposable incomes are now the group most likely to have access to broadband internet connections, MP3 players and next-generation mobile devices (Kuo 2007; CNNIC 2008).

The Chinese government has been reluctant to abandon cultural policies that place heavy emphasis on the pedagogical and political role of cultural activities. In spite of this, opportunities for commercially driven cultural industries are increasing (Liao 2006). However, while political sensitivities are still a factor, people are making and consuming music widely, and businesses are finding ways to generate income around these activities. Policies originally intended to control heterodox content have had another important effect: they have created barriers to the legitimate domestic market for foreign content producers, increasing incentives for the production of domestic content and reducing foreign competition. Although the structures that define China's commercial music industry are still crystallizing, it is already possible to see important differences between the business models and industry structures that evolved in the United States and those that are emerging in China.

One strategy for making money in the absence of strong copyright has been to rely on personal appearances by artists, which cannot be replicated. As a result, there is less emphasis on producing popular albums and more emphasis on gaining popularity and profile through single hits that lead to lucrative product endorsement and live appearance or performance deals (Wang 2005). However, even for Chinese labels, relying on personal appearance and advertising revenue presents practical problems, including limited scalability and continuing sensitivity over large popular music events (China Music Radar 2008). Furthermore, advertising and personal appearance are difficult to reconcile with the 'long tail' approach, which, in other markets, allows back-catalogues to continue generating revenue for labels and artists long after the artist has been eclipsed by the latest trend.

As a result, the distribution of music to mobile devices is quickly becoming one of the most significant sites of economic activity associated with music in China (Yao 2007). Just as analogue technologies allowed a limited number of firms in Europe and the United States to control the physical production and mass distribution of music for much of the twentieth century, mobile networks are making it possible for a few key players to control the distribution of content to mobile devices and the collection of payments for the use of mobile music services. In other markets, record labels emerged as the most powerful group in the Western recorded music industry, controlling access to capital, production of physical music products and distribution channels. In China, mobile operators are on track to play a similar role. The existence of a formal copyright law is having an impact on the strategies being employed by firms seeking to capitalize on consumer demand for music. However, the use of physical technologies for channelling access and managing micro-payment collection is proving far more influential.

All rent seeking thus involves, in some way or other, the tacit presumption that business models are parametric, like law. This is of course no less true in the creative industries than in other industries. Yet the creative industries represent an important challenge to this presumption because creative industries business models change so often as firms in this sector of the economy adapt to the existence of new technologies, seek to capitalize on new markets and alter their strategies to take advantage of developments in law and policy. Business models are not parameters about which law seeks to form and solidify, but rather continually adaptive technologies that take particular structures of law as aspects of the business environment. As a result, although strong intellectual property rights are unlikely to lead to growth in the creative industries, they may prevent firms seeking to capitalize on the opportunities of global creative industries markets from developing business strategies that will assist them in doing so. Strong competitive advantage in the creative industries therefore flows from weak, not strong, intellectual property environments.

CONCLUSION

This chapter has argued that important flaws in standard economic theories of intellectual property exist in relation to copyright. The limitations of economic models of intellectual property's effects are particularly apparent when the characteristics of the creative industries are considered. The creative industries operate in predominantly global markets. Their economic value is closely associated with the reuse of creative works and ideas and the transfer of knowledge within and across industries and domains. As a comparison of the commercial music industries in the United States and China demonstrates, creative industries business models co-evolve with legal and institutional environments. As a result, a weak intellectual property system is, in theory, capable of functioning as a competitive strength both for producers and for consumers in the creative industries.

The creative industries are a special but illustrative case, because their key properties – global context, knowledge reuse and adaptive business models – are properties that are likely to become more common throughout the economy as digital technologies and globalization continue to make an impact on business environments. The rapid development of the creative industries in China appears to contradict the hypothesis that stronger IP is the pathway to economic growth. Instead, this chapter argues that weaker IP is a much overlooked source of evolutionary development. As new technologies and globalization increase levels of connectivity among consumers and creators, approaches that are proving successful in China may well turn out to be at the vanguard of models for monetizing creativity in a digital age.

NOTES

1. This chapter is largely based on the journal article 'Does Weaker IP Mean Stronger Creative Industries? Some Lessons from China', with J. Potts, *Journal of Creative Industries* (2009), 1 (3), 245–61. As such, it owes a great deal to Jason Potts, senior research fellow at the Centre of Excellence for Creative Industries and Innovation, Queensland University of Technology.
2. An excellent discussion of this history can be found in Deazley (2006).
3. It is noteworthy that a parallel debate is occurring with the study of the economics of 'open-source' production and innovation (see Lerner and Tirole 2002).

7. Conclusion: transition phase or sign of things to come?

In this book I have argued that social network markets and consumer creativity and entrepreneurship are important forces in the production and commercialization of cultural commodities in the twenty-first century. I have also argued that spaces between policy and practice are functioning as key sites for the generation of new knowledge and the evolution of new approaches to the business of culture in a digital age. The global circulation of content, technological transformations and new possibilities for networked creation and distribution are helping to drive innovation and growth in China's creative industries. The same phenomena are also creating new challenges for businesses, policymakers, academics and consumers, all of whom are struggling to understand and navigate the complex, changing dynamics of creative production and consumption.

Since the establishment of the World Intellectual Property Organization (WIPO) in 1970, the breadth, scope and terms of global intellectual property protection have expanded steadily (Boyle 2004). As policymakers, businesses and users jostle to optimize their interests in a rapidly changing social, economic and technological landscape, debates about how the rights of those who invest in creative production should be protected and how the benefits to society of creativity and innovation might be maximized are taking on new significance. Increases in the value of intellectual property-related trade have been associated with pressure for developing nations to adopt higher standards of IPR protection (Drahos and Braithwaite 2002). In many instances access to the benefits of international trade is now conditional on protection for intellectual property rights (Maskus 2000).

Even the most optimistic assessments of the impact of new frameworks for the global regulation of intellectual property rights, such as TRIPs, acknowledge that the greatest burden of expanding global IP frameworks falls on developing nations, whose governments must bear the costs of implementing new laws and whose citizens are required to pay for information and content which would otherwise be freely available (Maskus 2000: 6). Understanding the extent to which intellectual property protection might enable the growth of domestic innovative and creative capacities is thus a key question for developing countries as they strive to meet their development goals (de Beer 2009).

In China's case, economic reform and a desire to gain access to global trading opportunities have been associated with the construction of an intellectual
property system that, on paper at least, satisfies the demands of major trading
partners, such as the United States and the European Union (Xue and Zheng
2002; Montgomery and Fitzgerald 2006). Considerable resources are being
devoted to training lawyers and judges, developing administrative mechanisms for the enforcement of intellectual property rights and establishing dedicated intellectual property courts. But, although progress is being made in the
construction of comprehensive legislative and judicial infrastructures for the
protection of IPRs, levels of enforcement remain low. As this book has
discussed, some estimates suggest that up to 90 per cent of film and music
products consumed in China are 'pirated' (MPAA 2004; Kennedy 2006),
enforcement of intellectual property rights in online environments remains a
serious challenge for copyright owners, and counterfeit versions of branded
consumer items are widely available.

At the same time, policymakers at national, provincial and city levels are
pushing forward with cultural sector reforms intended to promote the growth
of an 'innovative nation' (Chang 2009). The rhetoric of the creative industries
is gaining traction at provincial and city levels, and investment in the creative
industries is being encouraged (National Animation Industry Base n.d.; Keane
2009). These developments are taking place in the context of broader efforts
to reduce dependence on low-cost manufacturing as a source of employment
and growth and to develop higher-value-added areas of the economy (Chang
2009; Sheng 2009).

Some of the characteristics of the case studies explored in this book might
be understood as unique to China and its history. However, there are also more
general lessons to be taken from examining the ways in which innovation, the
role of the consumer and transformative technological change are interacting
with technologies for the governance and coordination of markets in creative
and cultural products. Although the creative industries include the 'copyright
industries', they are much larger than the copyright industries alone.
Furthermore, the speed with which digital technologies have revolutionized
creative practices, distribution systems and consumer expectations has been
breathtaking. It is no surprise, then, that there is tension between legal frameworks intended to help coordinate markets in creative and cultural works, such
as copyright, and the needs and aspirations of businesses, consumers and
governments faced with the new opportunities, and difficulties associated with
global interaction and digital affordance.

The creative industries are an area of the economy that is being profoundly
affected by transformative technological change. Technological innovation
and dissemination are bringing new products, services and markets into existence, and challenging the effectiveness of business models in established

industries, including the three explored in this book. In China, transformative physical technologies have become available in the context of sweeping economic reform processes and increased interaction with international communities of trade, as well as important changes in strategies for governing the cultural sector, and the development of new legal infrastructures available to the state, citizens and businesses. Furthermore, as Chapter 6 discussed, there is some evidence that a strong intellectual property system may not be the most significant factor in the emergence of commercially successful creative industries in China, where the film, mobile music and fashion industries are all developing rapidly in spite of a weak copyright environment.

Given the complexity of the environment within which creative businesses in China must operate, how might the role of intellectual property in promoting growth and innovation in the creative industries be understood? Evolutionary economics provides a framework through which the complex relationships between formal law, strategies for governance, physical technology and consumers might be conceptualized. As BOP Consulting points out:

> One of the most useful contributions that the evolving discipline of 'evolutionary economics' makes is the observation that neo-classical economics assumes that demand and supply exist in the first place. That is, it offers no account of how an economy comes into being. Further, by treating innovation as an exogenous process, neo-classical economics also has little to say about processes of industrial change and restructuring. (BOP Consulting 2010)

Neoclassical economics approaches innovation as something that occurs outside the economic model (exogenous) rather than as a function of dynamic processes occurring within the economic system itself. According to this model, innovation in physical technology produces growth in the economic system, which would otherwise tend towards an equilibrium point. But, as the above quote hints, the neoclassical model is unable to explain how technological innovations occur or how they are integrated into the economic system.

The neoclassical approach is able to explain how a change in a physical process, for example a more efficient method of producing fabric, contributes to economic growth in an industrial sector (in this case the textile industry) that supplies an existing market (such as garment manufacturers). Both producers and consumers in this context already understand what fabric might be used for. However, it is unable to explain how either businesses or consumers discover what a transformative technology might be used for, how the new products and services that such a technology makes possible are integrated into existing industry structures and consumer practices, or how the value of a previously unimagined technology, product or service is decided.

The limitations of the neoclassical approach are particularly apparent when attempting to understand the economics of the creative industries, which involve constant innovation, as well as trades in intangible products and services. A neoclassical model may tell us something about the role that intellectual property plays in industries with established value chains, business models and markets. However, it is not very helpful when it comes to explaining the relationship between intellectual property protection, innovation and the growth and coordination of new markets arising from transformative technological change in dynamic sectors such as the creative industries. Thus a growing body of research on the economics of the creative industries has turned to evolutionary approaches in an attempt to better explain how value is generated in the creative industries and how this sector of the economy relates to wider processes of growth and change in the economic system.

In contrast to neoclassical economists who view economic growth as a function of technological innovation, Beinhocker (2006) and other evolutionary economists argue that growth is a result of dynamic processes of change that involve physical technologies, social technologies and business models. Beinhocker argues that wealth creation is the product of the three-stage process that Darwin observed in *The Origin of Species*: differentiation, selection and amplification. Although evolution was first observed by Darwin in relation to biological phenomena, modern theorists view evolution as a process that occurs much more widely. As Beinhocker puts it: 'Evolution is an *algorithm*; it is an all-purpose formula for innovation, a formula that, through its special brand of trial and error, creates new designs and solves difficult problems' (Beinhocker 2006: 12).

Evolutionary economists argue that differentiation, selection and amplification drive wealth creation by acting as the mechanism through which effective solutions to economic problems emerge from the almost infinite number of less effective possibilities. Human creativity, skill and intentionality play an important role in this process, as individuals, organizations and societies devise solutions that they think may solve particular problems and then test them out in the environment in which they must function. However, the complexity of the economic challenges that human society faces means that trial and error also have a role to play:

> A variety of candidate designs are created and tried out in the environment; designs that are successful are retained, replicated, and built upon, while those that are unsuccessful are discarded. Through repetition, the process creates designs that are fit for their particular purpose and environment. (Beinhocker 2006: 14)

Although the mechanism through which designs are selected appears fairly simple, Beinhocker argues that economic evolution in fact involves three interlinked processes: the evolution of physical technologies, the evolution of social technologies and the evolution of business. The term 'physical technologies' describes things like arrow-head designs, the printing press and the microchip. Social technologies are methods and designs for organizing people, such as settled agriculture, political systems and the rule of law. Businesses make it possible for physical and social technologies to have an impact on the world, by translating innovations into new products or services (Beinhocker 2006: 16).

As an emerging field, evolutionary economics is not directly concerned with identifying specific solutions to the complex problems faced by human society. Rather, it is interested in describing and explaining mechanisms for change, adaptation and wealth creation in the economic system over time. In doing so, an evolutionary approach is valuable to those seeking to understand how new ideas and technologies are converted into economic benefit, as well as the impact of major changes in the environment within which particular economic activities occur.

The creative industries are part of just such a complex system. They are inextricably linked to both the broader economy and the complex landscapes of culture, identity, language, trade, law and morality in which they operate. Business models in the creative industries co-evolve with technological developments such as the internet, as well as social technologies, which include the dynamic technologies of governance described by Foucault and Yurchak, as well as the formal legal system and constructs such as copyright law. Thus it seems sensible that any attempt to understand the role that intellectual property plays in promoting economic growth in the creative industries should take into account the dynamic nature of the physical technologies available to businesses and consumers operating in the creative industries, the capacity of social technologies to change, and the co-evolution of business models alongside these changing systems.

The dynamic co-evolution of physical technologies, social technologies and business models is evident in the three case studies explored in this book. Social technologies, including strategies for governance, are evolving in China in response to changes in physical technology that have challenged the effectiveness of attempts to directly control the production and consumption of cultural products and made it possible for consumers to become more actively involved in processes of creation and distribution. The broader economic environment within which both cultural production and consumption are occurring has also altered, as a result of the adoption of market-oriented economic policies. Creative industries businesses are being given space to experiment with new business models, and in so doing they are identifying those capable of functioning within the rules of the Chinese system. Although copyright law exists, it is just one factor among many influencing the behaviour of both businesses and consumers.

Film continues to be viewed by China's policymakers as an important site of popular education, as well as entertainment and commercial opportunity.

Government policies are encouraging entrepreneurs to produce images of individuality, morality and nationhood deemed suitable for mass consumption by
the state. Although copyright protection remains problematic and audiences
have access to content through the internet and unauthorized DVD distribution
networks, private investment in film production and the nation's cinema
networks is being encouraged within careful limits. There can be no doubt that
access to a coordinated cinema distribution system is providing film industry
entrepreneurs and investors with valuable space in which to operate. The
controlled mass distribution system provided by China's cinema networks is
making large-scale investment in Chinese film production both conceivable
and profitable.

Certain aspects of the music industry, on the other hand, enjoy much
higher levels of autonomy in relation to the nature of the content they
produce and circulate. The possibility of removing politically sensitive
material from mass circulation at a handful of centralized distribution points
is real, and international artists face higher levels of scrutiny than their local
counterparts. This power is helping to limit international competition in the
domestic mobile music market and is arguably providing Chinese content
producers with a powerful advantage. As with the film industry, access to a
closed distribution network is playing a vital role in the coordination of
China's music industry. It appears that mobile technology is creating a distribution monopoly that, in other markets, has been achieved through the
enforcement of copyright law.

In contrast to both film and music, fashion in China has remained an area
in which distributed production and individual agency in consumption
choices have been the norm, even during the most ideologically extreme
periods of the nation's recent history (Finnane 2007). The nature of fashion
consumption, distribution and the creative and value-generating processes of
design and brand establishment is such that production and consumption are
both dynamic and discursive, closely connected to society-wide changes in
attitudes and values. Thus Chinese consumers are playing an essential role
in the emergence of a Chinese fashion industry that is quickly moving
beyond its role as the low-cost manufacturer of foreign designs. As in other
markets, China's fashion industry is being driven by investments in reputation by designers and consumers, and relies heavily on the branch of intellectual property concerned with maintaining a relationship between a
product and its origin: trademarks. Although controlled distribution of legitimate versions of trademarked products is important for the industry, the
widespread availability of 'fake' branded products does not appear to be
preventing the rapid growth in demand for genuine products.

COPYRIGHT: AN EVOLVING SOCIAL TECHNOLOGY

Copyright law is a social technology that, in some circumstances, helps to coordinate markets in creative expression. By creating a monopoly over the right to decide how a work is distributed and used, copyright helps to make scalable returns on investments in production, promotion and distribution possible. In the context of particular states of technological development, economic policies and systems of government, copyright has been instrumental in attracting investment into risky, capital-intensive creative activities such as the production of blockbuster films. It is difficult to imagine a more complex task for any legal mechanism. It is little wonder, then, that the history of intellectual property law is one of contestation and debate. That an intellectual property system capable of functioning well enough to allow global markets in popular films, music, books, radio, television and branded fashion items ever came into existence at all is in itself a breathtaking achievement.

However, the three case studies explored in this book make it clear that copyright's effectiveness as a coordinating mechanism depends heavily on factors that include systems of governance and developments in physical technology, the nature of the product an industry trades in and the dynamics of its consumption. Furthermore, copyright law is just one of a number of possible solutions to the tricky problem of how a market in creative products might be coordinated. China's film and music industries are able to rely on distribution monopolies provided by access to state-controlled distribution networks, such as cinema, as well as technologically closed distribution systems such as mobile networks. The extent to which different sectors of the creative industries depend on copyright protection and the distribution monopoly that it provides varies enormously. China's fashion industry, like fashion industries elsewhere in the world, depends on trademarks, rather than copyright, and relies on active consumers who are involved in productive processes, rather than linear models of product development.

Cunningham *et al.* (2009) argue that the creative industries are best understood as part of an innovation system, rather than as an industrial sector in their own right:

> the creative industries are not an industrial sector in the sense that they produce a particular set of commodities (as compared with the coal industry, for example), but, rather, are essentially engaged in the provision of coordination services that relate to the origination, adoption and retention of new technologies, commodities or ideas into the economic system. They provide, in other words 'evolutionary or innovation services'. (Cunningham *et al.* 2009: 6)

I would argue that entrepreneurial consumers must also be considered as part of the innovation system, as a potentially powerful force in dynamic processes of market coordination and wealth creation.

Of the three case studies explored in this book, the role of entrepreneurial consumers in processes of origination, adoption and retention is most obvious in fashion. Entrepreneurial consumers are making it possible for China's fashion industry to grow and compete in an international market for designs and brands. The industry's capacity to shift from low-cost manufacturer of the designs of others to designer of trends adopted around the world is closely connected to consumer familiarity with the rules of the international fashion system, and willingness to adopt the products and designs offered to them by Chinese designers. Although designers are able to provide consumers with products that they hope will prove popular, it is ultimately consumers who decide whether or not trends are adopted, what represents value and whether the image and identity that have been cultivated by a particular brand warrant the purchase price demanded.

However, the fact remains that, while the fashion system has been adept at accommodating the innovative and entrepreneurial tendencies of consumers within coordinated systems of mass production and consumption, commercially successful elements of China's film and music industries continue to depend heavily on an ability to control distribution and use of the creative products that they trade in. Although copyright law has not, so far, provided the basis for this control, it is taking time for business models capable of functioning where such control does not exist to emerge. How entrepreneurial consumers might be more comfortably incorporated into the productive processes of commercial film and music industries in a digital age remains a challenge in all markets.

Although China has many unique characteristics, the challenges faced by its rapidly developing creative industries also have a great deal in common with the challenges that are confronting creative industries elsewhere in the world. The struggle to find business models capable of capitalizing on creative populations, peer-to-peer communication and transformative technological change are precisely the challenges facing established copyright industries in the United States, the UK and Australia. Recognizing the dynamic characteristics of the creative industries and the capacity of physical technology, legal structures and business models to change and adapt will be vital to ensuring that intellectual property theory, and policy, is able to respond to these challenges effectively.

This book has explored changing dynamics of cultural production and consumption, and considered the role that intellectual property and entrepreneurial consumers are playing in the development of three very different creative industries in China. It seems that what may be needed by twenty-first-century creative industries, and creative consumers, may involve a hybrid approach which is able to combine the most useful elements of copyright's focus on rewarding creators and investors in creative works with the fashion system's ability to capture the economic value of reputation, and to be driven forward by consumers who are actively engaged in creative processes.

References

Alford, William (1995), *To Steal a Book Is an Elegant Offense*, Stanford, CA: Stanford University Press.

Allen Consulting Group (2001), 'The Economic Contribution of Australia's Copyright Industries', available at www.copyright.org.au/publications/research/bcepv03.pdf (accessed 3 February 2010).

Arup, Christopher (2000), *The New World Trade Organization Agreements: Globalizing Law through Services and Intellectual Property*, Cambridge: Cambridge University Press.

Au, Kin-fan (2009), 'The Maturing Apparel Industry in China', available at http://www.udel.edu/fiber/issue3/world/ApparelIndustry.html (accessed 3 February 2010).

Baranovitch, Nimrod (2003), *China's New Voices: Popular Music, Ethnicity, Gender and Politics*, Berkeley: University of California Press.

Barnett, Jonathan (2005), 'Shopping for Gucci on Canal Street: Reflections on Status Consumption, Intellectual Property and the Incentive Thesis', *Virginia Law Review*, 91 (6) [Online], available at http://www.virginialawreview.org/articles.php?article=76 (accessed 3 February 2010).

Becker, Jasper (2000), *The Chinese*, London: John Murray Publishers.

Beinhocker, Eric (2006), *The Origin of Wealth*, New York: Random House.

Bettig, Ronald (1996), *Copyrighting Culture: The Political Economy of Intellectual Property*, Boulder, CO: Westview Press.

Boldrin, Michele and David K. Levine (2002), 'The Case against Intellectual Property', *American Economic Review: Papers and Proceedings*, 92 (2), 209–12.

Boldrin, Michele and David K. Levine (2005), 'The Economics of Ideas and Intellectual Property', *Proceedings of the National Academy of Sciences*, 102, 1252–6.

Boldrin, Michele and David K. Levine (2008), *Against Intellectual Monopoly*, Cambridge: Cambridge University Press.

BOP Consulting (2010), *Literature Review: Changing Attitudes and Behaviour in the 'Non-Internet' Digital World and Their Implications for Intellectual Property*, London: Strategic Advisory Board on Intellectual Property.

Boyle, James (2004), 'A Manifesto on WIPO and the Future of Intellectual Property', *Duke Law and Technology Review*, 9, 1–12.

Brace, Tim (1991), 'Popular Music in Contemporary Beijing: Modernism and Cultural Identity', *Asian Music*, 22 (2), 43–66.

Brady, Anne-Marie (2006), 'Guiding Hand: The Role of the CCP Central Propaganda Department in the Current Era', *Westminster Papers in Communication and Culture*, 3 (1), 57–76.

Breward, Christopher (2000), 'Cultures, Identities, Histories: Fashioning a Cultural Approach to Dress', in Nicola White and Ian Griffiths (eds), *The Fashion Business: Theory, Practice, Image*, Oxford and New York: Berg, pp. 23–36.

Breward, Christopher (2003), *Fashion*, Oxford: Oxford University Press.

Bruns, Axel (2008), *Blogs, Wikipedia, Second Life, and Beyond: From Production to Produsage*, New York: Peter Lang Publishing.

Burchell, Graham (1996), 'Liberal Government and Techniques of the Self', in Andrew Barry, Thomas Osborne and Nikolas Rose (eds), *Foucault and Political Reason: Liberalism, Neo-liberalism and Rationalities of Government*, London: University College London Press.

Burnett, Robert (1996), *The Global Jukebox: The International Music Industry*, London: Routledge.

Centre for Cultural Policy Research (2003), *Baseline Study on Hong Kong's Creative Industries for the Central Policy Unit, Hong Kong Special Administrative Region Government*, Hong Kong: University of Hong Kong.

Chang, Ha-Joon (2008), *Bad Samaritans: The Myth of Free Trade and the Secret History of Capitalism*, London: Random House Business Books.

Chang, Shaun (2009), 'Great Expectations: China's Cultural Industry and a Case Study of a Government-Sponsored Creative Cluster', *Creative Industries Journal*, 1 (3), 263–73.

Chen, Jianfu (1999), *Chinese Law: Towards an Understanding of Chinese Law, Its Nature and Development*, The Hague: Martinus Nijhoff Publishers.

China Daily (2009), 'Copyright Centre Up and Running', available at http://www.chinaipr.gov.cn/news/enterprise/251165.shtml (accessed 10 February 2010).

China Economic Review (2004), 'Regulator Monster Biffs Spiderman', available at http://www.chinaeconomicreview.com/cer/2004_08/Regulator_Monster_biffs_Spiderman.html (accessed 2 February 2010).

China Film Press (1995–2007), *China Film Yearbook*, Beijing: China Film Press.

China Mobile (2007), 'Interim Results', available at www.chinamobileltd.com/images/pdf/2007/ir_2007_e.pdf (accessed 3 February 2010).

China Mobile (2008), *FAQ*, available at http://www.chinamobileltd.com/ir.php?menu=5 (accessed 11 May 2008).

China Music Radar (2008), 'MIDI Festival Cancelled', *China Music Radar*

[Online], 24 April, available at http://www.chinamusicradar.com/?p=15 (accessed 7 February 2010).

China Research Intelligence (2009), 'Report of Chinese Apparel Industry, 2009', available at http://www.thefreelibrary.com/Report+of+Chinese+Apparel+Industry,+2009-a01073945145 (accessed 8 February 2010).

China Retail News (2008), 'Ministry of Commerce: China Will Become World's Biggest Luxury Market by 2014', *China Retail News* [Online], 18 April, available at http://www.chinaretailnews.com/2008/04/18/1134-ministry-of-commerce-china-will-become-worlds-biggest-luxury-market-by-2014/ (accessed 4 February 2010).

China Tech News (2006), 'Hurray! Joins Hands with MTV', available at http://www.chinatechnews.com/2006/07/11/4260-hurray-joins-hands-with-mtv/ (accessed 4 February 2010).

China Unicom (2008), *Company Profile*, http://www.chinaunicom.com.hk/en/aboutus/profile.html (accessed 22 February 2008).

Chow, Daniel (2006), 'Counterfeiting and China's Economic Development', *US–China Economic and Security Review Commission*, 8 June, available at http://www.uscc.gov/hearings/2006hearings/written_testimonies/06_06_08wrts/06_06_7_8_chow_daniel.pdf (accessed 4 February 2010).

CMM Intelligence (2004), 'US Film Imports to Be Cut? SARFT: Yes . . . No . . . Maybe, I Don't Know . . . Can You Repeat the Question?', *CMM Intelligence*, 8 (4) [Online], available at http://www.cmmintelligence.com/?q=node/4510.

CNNIC (2008), *Statistical Survey Report on the Internet Development in China (January 2008)*, available at http://www.cnnic.cn/uploadfiles/pdf/2008/2/29/104126.pdf (accessed 25 July 2008).

CNNIC (2009), *Statistical Survey Report on the Internet Development in China (July 2009)*, available at http://www.cnnic.net.cn/uploadfiles/pdf/2009/10/13/94556.pdf (accessed 4 February 2010).

Cohen, Wesley, Richard Nelson and John Walsh (2000), 'Protecting Their Intellectual Assets: Appropriability Conditions and Why U.S. Manufacturing Firms Patent (or Not)', National Bureau of Economic Research Working Paper no. W7552, available at http://papers.ssrn.com/sol3/papers.cfm?abstract_id=214952 (accessed 4 February 2010).

Colvin, Mark (2006), 'China's Leaders "Riding the Tiger": Expert', *ABC Radio* [Radio broadcast], 9 August, available at http://www.abc.net.au/pm/content/2006/s1711015.htm (accessed 8 February 2010).

Consumers International Asia Pacific Office (2006), *Copyright and Access to Knowledge: Policy Recommendations on Flexibilities in Copyright Laws*, Kuala Lumpur: Consumers International.

Coonan, Clifford (2008), 'China Tightens Rules on Foreign Performers', *Variety* [Online], 18 July, available at http://www.variety.com/article/VR1117989152.html?categoryid=16&cs=1 (accessed 4 February 2010).

Cox, Christine and Jennifer Jenkins (2005), 'Between the Seams, a Fertile Commons: An Overview of the Relationship between Fashion and Intellectual Property', available at http://learcenter.org/pdf/RTSJenkins Cox.pdf (accessed 3 February 2010).

Credit Suisse Equity Research (2005), *China Internet Sector: Mobile Music Revolution in China*, Sector Review, Zurich: Credit Suisse.

Crofton, I. (1988), *A Dictionary of Art Quotations*, London: Routledge.

Cunningham, Stuart (2006), *What Price a Creative Economy?*, Strawberry Hills, NSW: Currency House.

Cunningham, Stuart, John Hartley, Jason Potts, Arthur ter Hofstede, Julian Thomas, Denise Meredyth, Ellie Rennie and Brian Fitzgerald (2009), 'Submission to the Review of the National Innovation System', available at http://docs.google.com/viewer?a=v&q=cache:pW1_CskPJ dEJ:www. innovation.gov.au/innovationreview/Documents/261-ARC-CCI.pdf+ Cunningham+Potts,+National+Innovation+Submission &hl= en&gl= uk&sig=AHIEtbTn1uE0wITiIjsofQO7eMvGK0PkOQ (accessed 5 February 2010).

Daniel, Mathew (2007), 'So You Want to Sell Music in China?', available at http://outdustry.com/2008/01/17/so-you-want-to-sell-music-in-china-guest-post/ (accessed 4 February 2010).

DCMS (1998), *Creative Industries Mapping Document*, London: HMSO.

DCMS (2010), 'Creative Industries', available at http://www.culture.gov.uk/ about_us/creative_industries/default.aspx (accessed 4 February 2010).

Deazley, Ronan (2006), *Rethinking Copyright: History, Theory, Language*, Cheltenham, UK and Northampton, MA, USA: Edward Elgar Publishing.

de Beer, Jeremy (2009), 'Defining WIPO's Development Agenda', in Jeremy de Beer (ed.), *Implementing the World Intellectual Property Organization's Development Agenda*, Waterloo, ON: Wilfrid Laurier University Press.

de Kloet, Jeroen (2002), 'Rock in a Hard Place: Commercial Fantasies in China's Music Industry', in Stephanie Donald and Michael Keane (eds), *Media in China: Consumption, Content and Crisis*, London: RoutledgeCurzon, pp. 93–104.

Dixon, Padraig and Christine Greenhalgh (2002), 'The Economics of Intellectual Property: A Review to Identify Future Research Directions', Nuffield College Working Paper no. 0502 [Online], available at www.oiprc.ox.ac.uk/EJWP0502.pdf (accessed 4 February 2010).

Dopfer, Kurt and Jason Potts (2004), 'Evolutionary Foundations of Economics', in J. Stanley Metcalfe and John Foster (eds), *Evolution and Economic Complexity*, Cheltenham, UK and Northampton, MA, USA: Edward Elgar Publishing.

Dopfer, Kurt and Jason Potts (2008), *The General Theory of Economic Evolution*, London: Routledge.

Drahos, Peter and John Braithwaite (2002), *Information Feudalism: Who Owns the Knowledge Economy?*, London: Earthscan Publications.

Economist (2006), 'No Direction: Everyone Is in Love with Chinese Cinema. Except the Chinese' [Print edition], 27 April.

Endeshaw, Assafa (1999), *Intellectual Property in China: The Roots of the Problem of Enforcement*, Singapore: Acumen Publishing.

European Audiovisual Observatory (2009), *Marché du Film 2009: World Film Market Trends*, available at http://www.international-television.org/tv_market_data/focus_world_film_market_trends_statistics.html (accessed 11 May 2010).

European Commission (2008), '2007 Customs Seizures of Counterfeit Goods – Frequently Asked Questions', Press Release [Online], 19 May, available at http://europa.eu/rapid/pressReleasesAction.do?reference=MEMO/08/310 (accessed 4 February 2010).

Ewan, Elizabeth (1990), *Townlife in Fourteenth Century Scotland*, Edinburgh: Edinburgh University Press.

Fine, Jon (2007), 'Opinion: Leaving Record Labels Behind', *Business Week* [Online], available at http://www.businessweek.com/magazine/content/07_44/b4056094.htm (accessed 7 February 2010).

Farrell, Diana, Ulrich Gersch and Elizabeth Stephenson (2006), 'The Value of China's Emerging Middle Class', *McKinsey Quarterly*, 2006 special edition: *Serving the New Chinese Consumer*, New York: McKinsey & Company, pp. 60–69.

Finnane, Antonia (2007), *Changing Clothes in China: Fashion, History, Nation*, London: Hurst Publishers.

Fisher, Marshall, Janice Hammond, Walter Obermeyer and Annath Raman (1994), 'Making Supply Meet Demand in an Uncertain World', *Harvard Business Review*, May–June, 83–93.

Fitzgerald, Anne and Brian Fitzgerald (eds) (2004), *Intellectual Property in Principle*, Sydney: Lawbook Company.

Foucault, Michel ([1961] 1989), *Madness and Civilization: A History of Insanity in the Age of Reason*, translated from French by Richard Howard, London and New York: Routledge.

Foucault, Michel (1972), *The Archaeology of Knowledge*, translated from French by S. Smith, London: Routledge.

Foucault, Michel (1978), *The History of Sexuality*, Vol. 1: *An Introduction*, translated from French by R. Hurley, Harmondsworth: Penguin.

Foucault, M. (1984), 'What Is Enlightenment?', in *The Foucault Reader*, ed. P. Rabinow, New York: Pantheon.

Foucault, Michel (1985), *The History of Sexuality*, Vol. II: *The Use of Pleasure*, translated by R. Hurley, London: Penguin.

Foucault, Michel (1986), *The History of Sexuality*, Vol. III: *The Care of the Self*, translated by R. Hurley, London: Penguin.

Foucault, Michel (1988), 'Technologies of the Self', in Patrick Hutton (ed.), *Technologies of the Self: A Seminar with Michel Foucault*, Amherst: University of Massachusetts Press.

Foucault, Michel (2002), *Power: Essential Works of Foucault 1954–1984*, translated from French by Robert Hurley, ed. James Faubion, London: Penguin Books.

Frith, Simon (2004), 'Copyright, Politics and the International Music Industry', in Simon Frith and Lee Marshall (eds), *Music and Copyright*, Edinburgh: Edinburgh University Press, pp. 70–88.

Gilbert, Richard and Carl Shapiro (1990), 'Optimal Patent Length and Breadth', *RAND Journal of Economics*, 21, 106–12.

Ginsburg, Jane (1992), 'Moral Rights in a Common Law System', in Peter Anderson and David Saunders (eds), *Moral Rights Protection in a Copyright System*, Brisbane: Griffith University Press.

Grinvald, Leah (2008), 'Making Much Ado about Theory: The Chinese Trademark Law', *Michigan Telecommunications and Technology Law Review*, 15 (53), 53–106.

Gronow, Pekka and Ilpo Saunio (1999), *International History of the Recording Industry*, translated from Finnish by Christopher Moseley, London: Continuum International.

Halbert, Debora (1999), *Intellectual Property in the Information Age: The Politics of Expanding Ownership Rights*, London: Quorum Books.

Hancock, Terry (2006), 'The Case for a Creative Commons "Sunset" Non-Commercial Licence Module', *Free Software Magazine* [Online], 31 May, available at www.freesoftwaremagazine.com/node/1566 (accessed 2 February 2010).

Hartley, John (2005), 'Creative Industries', in John Hartley (ed.), *Creative Industries*, Oxford: Blackwell, pp. 1–40.

Hartley, John (2009), *The Uses of Digital Literacy*, Saint Lucia: University of Queensland Press.

Hartley, John and Lucy Montgomery (2009), 'Fashion as Consumer Entrepreneurship: Emergent Risk Culture, Social Network Markets, and the Launch of *Vogue* China', *Chinese Journal of Communication*, 2 (1), 61–76.

Hill, Daniel D. (2004), *As Seen in Vogue: A Century of American Fashion in Advertising*, Lubbock: Texas University Press.

Hindess, Barry (1996), *Discourses of Power: From Hobbs to Foucault*, Oxford: Blackwell.

Hirshleifer, Jack (1971), 'The Private and Social Value of Information and the Reward to Inventive Activity', *American Economic Review*, 61, 561–74.

HKTDC (2008), 'China's Luxury Consumption Market Heating Up', available at http://www.hktdc.com/info/vp/a/tjo/en/1/3/1/1X003GM2/China-S-

Luxury-Consumption-Market-Heating-Up.htm (accessed 4 February 2010).

Hooper, Beverley (1994), 'From Mao to Madonna: Sources on Contemporary Chinese Culture', *Southeast Asian Journal of Social Science*, 22, 161–9.

Hooper, Beverley (1998), 'Flower Vase and Housewife: Women and Consumerism in Post-Mao China', in Krishna Sen and Maila Stivens (eds), *Gender and Power in Affluent Asia*, Routledge: London, pp. 167–93.

Howkins, John (2001), *The Creative Economy*, London: Penguin.

Howkins, John (2005), 'Creativity, Innovation and Intellectual Property: A New Approach for the 21st Century', Paper presented at the 2005 Shanghai International IPR Forum, Shanghai, 2 December.

Hua, Jin (2004), 'Strengthen the Capital Market and Increase the Drive of the Cultural Industry', Paper presented to 2nd Annual Cultural Industries Conference, Taiyuan, Shanxi Province, 12–15 September.

Hui, Desmond (ed.) (2006), *Study on the Relationship between Hong Kong's Cultural and Creative Industries and the Pearl River Delta*, Hong Kong: Centre for Cultural Policy Research, University of Hong Kong.

Hui, Sylvia (2005), 'Massive Support Vowed for Creative Industries', *The Standard* [Online], 13 January, available at http://www.thestandard.com.hk/stdn/std/Front_Page/GA13Aa08.html (accessed 3 February 2010).

Hunt, Ken (2007), 'Welcome to the Radiohead Economy', *Globe and Mail* [Online], 29 November, available at http://www.theglobeandmail.com/report-on-business/article799838.ece (accessed 3 February 2010).

Hurray (2008), 'Company Introduction', available at http://www.hurray.com.cn/english/about_intro.htm (accessed 4 February 2010).

Ikels, Charlotte (1996), *The Return of the God of Wealth: The Transition to a Market Economy in Urban China*, Stanford, CA: Stanford University Press.

International Federation of Phonographic Industries (2008), 'Recording Industry Steps Up Its Campaign against Internet Piracy in China', available at http://www.ifpi.org/content/section_news/20080204.html (accessed 11 May 2010).

International Federation of the Periodical Press (2006), 'China', *FIPP World Magazine Trends 2006/7* [Online], available at http://www.fipp.com/News.aspx?PageIndex=2002&ItemId=13502 (accessed 3 February 2010).

International Telecommunications Union (2008), *The China and Hong Kong SAR ICT Markets*, available at http://www.itu.int/world2006/forum/china_hong_kong_sar.html (accessed 11 May 2010).

Jayasuriya, Kanishka (2001), 'The Rule of Law and Governance in East Asia', in Mark Beeson (ed.), *Reconfiguring East Asia: Regional Institutions and Organisations after the Crisis*, London: Curzon, pp. 99–116.

Jia, Wei (1998), 'China's Film Industry: Crisis or Transition', Ph.D. thesis, Australian National University, Canberra.

Keane, Michael (2007), *Created in China: The Great New Leap Forward*, London: Routledge.

Keane, Michael (2009), 'Understanding the Creative Economy: A Tale of Two Cities' Clusters', *Creative Industries Journal*, 1 (3), 211–26.

Keane, Michael and Stephanie Hemelryk Donald (2002), 'Responses to Crisis: Convergence, Content Industries and Media Governance', in Stephanie Hemelryk Donald and Michael Keane (eds), *Media in China: Consumption, Content and Crisis*, London: RoutledgeCurzon, pp. 200–211.

Kennedy, John (2006), 'Unlocking the Music Market in China', Speech delivered at the China International Forum on the Audio Visual Industry, Shanghai, 25 May, available at http://www.ifpi.org/content/section_views/view020.html (accessed 11 May 2010).

Kraus, Richard (2004), *The Party and the Arty in China: The New Politics of Culture*, Oxford: Rowman & Littlefield.

Kwan, Chi Hung (2008), 'Accelerating Appreciation of the RMB – A Major Step toward a Free Floating Exchange Rate System', *China in Transition* [Online], 8 May, available at http://www.rieti.go.jp/en/china/08050801.html (accessed 3 February 2010).

Kynge, James (2007), *China Shakes the World: The Rise of the Hungry Nation*, London: Phoenix.

Laing, Dave (2002), 'Copyright as a Component of the Music Industry', in Michael Talbot (ed.), *The Business of Music*, Liverpool: Liverpool University Press, pp. 171–94.

Lancaster, A. (2001), 'Hollywood vs China', *City Weekend*, 20 December.

Lanham, Richard (2006), *The Economics of Attention*, Chicago: University of Chicago Press.

Law, Ryan (2004), 'Cross-Border Traffic Co-productions between Hong Kong and Mainland China', *Far East Film* [Online], available at http://194.21.179.166/cecudine/datahost/fef2004/english/hongkong2004_3.html (accessed 22 October 2004).

Lemire, B. and G. Iello (2008), 'East and West: Textiles and Fashion in Early Modern Europe', *Journal of Social History*, 41 (4).

Lerner, Josh and Jean Tirole (2002), 'Some Simple Economics of Open Source', *Journal of Industrial Economics*, 50, 197–234.

Lessig, L. (2001), *The Future of Ideas: The Fate of the Commons in a Connected World*, New York: Random House.

Lessig, Lawrence (2004), *Free Culture: How Big Media Uses Technology and the Law to Lock Down Culture and Control Creativity*, New York: Penguin Press.

Liao, Han-Teng (2006), 'Towards Creative Da-Tong: An Alternative Notion of Creative Industries for China', *International Journal of Cultural Studies*, 9 (3), 395–406.

Liebowitz, Stan (1985), 'Copying and Indirect Appropriability: Photocopying of Journals', *Journal of Political Economy*, 95 (5), 945–57.

Lipsitz, George (1994), 'Who'll Stop the Rain? Youth Culture, Rock 'n' Roll, and Social Crises', in David Farber (ed.), *The 60's: From Memory to History*, Chapel Hill: University of North Carolina Press, pp. 206–34.

Litman, Jessica (2001), *Digital Copyright*, Amherst, NY: Prometheus Books.

Liu, Deming (2006), 'The Transplant Effect of Chinese Patent Law', *Chinese Journal of International Law*, 5 (3), 733–52.

Luppino, T. (2001), 'The True Art of a Copy Cat', *Cloudband Magazine* [Online], http://www.cloudband.com/magazine.articles2q01/exh_luppino_copying_050h.html (accessed 25 May 2005).

Marshall, Lee (2005), *Bootlegging: Romanticism and Copyright in the Music Industry*, London: Sage.

Maskus, Keith (2000), *Intellectual Property Rights in the Global Economy*, Washington, DC: Institute for International Economics.

McCormack, Richard (2006), 'China Replaces US as World's Largest Apparel Exporter: Trade Imbalances Could Cause Financial Upheaval; MAPI Analyst Implores U.S., IMF to Act Now on China's Yuan', *Manufacturing and Technology News*, 13 (16) [Online], available at http://www.manufacturingnews.com/news/06/0905/art1.html (accessed 3 February 2010).

McCullagh, Charles (2005), 'China: Clarification of Media Investment Regulations', *Magazine Publishers of America* [Online], available at http://www.magazine.org/international/13175.aspx (accessed 11 May 2010).

Mertha, Andrew (2005), *The Politics of Piracy: Intellectual Property in Contemporary China*, New York: Cornell University Press.

Miller, Toby, Nitin Govil, John McMurria, Richard Maxwell and Ting Wang (2005), *Global Hollywood 2*, London: British Film Institute.

Mills, Sara (2003), *Michel Foucault*, London: Routledge.

Ministry of Information and Industry (2008), *March 2008 Telecommunications Industry Statistical Report*, http://www.mii.gov.cn/art/2008/04/25/art_54_37170.html (accessed 8 March 2008).

Moller, Erik (2005), 'Are Creative Commons-NC Licenses Harmful?', *Podcasting News*, [Online], 4 October, available at http://www.podcastingnews.com/archives/2005/10/are_creative_co.html (accessed 3 February 2010).

Montgomery, Lucy and Brian Fitzgerald (2006), 'Copyright and the Creative Industries in China', *International Journal of Cultural Studies*, 9 (3), 407–18.

MPAA (2004), 'MPAA Calls on USTR to Use Special 301 to Leverage Pakistan, Russia, Taiwan, Malaysia and China to Reduce Piracy', available at www.ftac.net/6-MPAA_and_301.pdf (accessed 4 February 2010).

Murphy, Kevin, Benjamin Klein and Andres Lerner (2002), 'Intellectual Property: Do We Need It? The Economics of Copyright Fair Use in a Networked World', *American Economic Review: Papers and Proceedings*, 92 (2), 205–8.

Music 2.0: Exploring Chaos in Digital Music (2008), 'iTuneless iPod Faces Music Piracy in China', available at http://www.music2doto.com/archives/111 (accessed 15 December 2008).

National Animation Industry Base (n.d.), *National Animation Industry Base*, Wuxi: National Animation Industry Base.

National Bureau of Statistics of China (2005), *China Statistical Yearbook*, Beijing: National Bureau of Statistics of China.

O'Connor, Justin and Gu Xin (2006), 'A New Modernity? The Arrival of "Creative Industries" in China', *International Journal of Cultural Studies*, 9 (3), 271–83.

Office of the United States Trade Representative (2009), *2009 Special 301 Report* [Online], 30 April, available at http://www.mcit.gov.eg/General/IPR%20Report%202009%20by%20the%20Office%20of%20the%20United%20States%20Trade%20Representative2009517161614.pdf (accessed 4 February 2010).

Okonkwo, Uki (2007), *Luxury Fashion Branding*, Houndmills, Hampshire: Palgrave Macmillan.

Paloczi-Horvath, George (1963), *Mao Tse Tung: Emperor of the Blue Ants*, Garden City, NY: Doubleday.

Peerenboom, Randall (2002), *China's Long March towards Rule of Law*, Cambridge: Cambridge University Press.

Peto, Ed (2007), 'Enter the Dragon: Introduction to the Music Business in China', *China.Music* [Online], available at http://outdustry.com/2007/11/05/enter-the-dragon-introduction-to-the-music-business-in-china/ (accessed 3 February 2010).

Phillips, Jeremy and Alison Firth (2001), *Introduction to Intellectual Property Law*, London: Butterworths LexisNexis.

Potts, Jason (2003), 'Evolutionary Economics: An Introduction to the Foundation of Liberal Economic Philosophy', Discussion Papers Series no. 324, University of Queensland School of Economics, Brisbane.

Potts, Jason and Stuart Cunningham (2008), 'Four Models of Creative Industries', *International Journal of Cultural Policy*, 14 (3), 133–47.

Potts, Jason, Stuart Cunningham, John Hartley and Paul Ormerod (2008), 'Social Network Markets: A New Definition of the Creative Industries', *Journal of Cultural Economics*, 32 (3), 167–85.

Qu, Sanqiang (2002), *Chinese Copyright Law*, Beijing: Foreign Languages Press.

Romer, Paul (2002), 'When Should We Use Intellectual Property Rights?', *American Economic Review: Papers and Proceedings*, 92 (2), 213–16.

Royal Society for the Arts (2005), *Adelphi Charter on Creativity, Innovation and Intellectual Property*, available http://www.adelphicharter.org/adelphicharter.asp (accessed 3 February 2010).

Ruggieri, Maria (2002), 'Market Forces: Chinese Cinema in 2002', *Asian Film Connection* [Online], available at http://www.asianfilms.org/edres_links_detail.php?regionid=1&countryid=1&edlinkid=31%20&PHPSESSID=fd9d0 (accessed 3 February 2010).

Sauvé, Pierre, Olivier Barlet, Emmanuel Cocq, Madanmohan Rao, German Rey and Craig Van Grasstek (2006), 'Trends in Audiovisual Markets: China, Mongolia and South Korea', available at http://portal.unesco.org/ci/en/ev.php-URL_ID=22361&URL_DO=DO_TOPIC&URL_SECTION=201.html (accessed 15 May 2010).

Scott, Ajax (2005), 'UK Companies Aim to Crack Fast-Changing Chinese Market', *Music Week* [Online], 17 December, available at http://goliath.ecnext.com/coms2/gi_0199-5093751/UK-companies-aim-to-crack.html (accessed 3 February 2010).

Seidman, Anne and Robert Seidman (1994), *State and Law in the Development Process: Problem-Solving and Institutional Change in the Third World*, New York: St. Martin's Press.

Selden, Mark (1995), *China in Revolution: The Yenan Way Revisited*, New York and London: M.E. Sharpe.

Sheng, Lu (2009), 'The Outlook for U.S.–China Textile and Apparel Trade in 2009: From the Trade Policy Perspective', *Fiber*, 3, available at http://www.udel.edu/fiber/issue3/world/ApparelTradeOutlook.html (accessed 3 February 2010).

State Administration of Radio, Film and Television (2003), 'Interim Provisions on the Access of Operational Qualifications for Movie Production, Distribution and Projection', *Laws of the People's Republic of China* [Online], 29 October, available at http://www.asianlii.org/cn/legis/cen/laws/ipotaooqfmpdap1103/ (accessed 7 February 2010).

Steele, Valerie (2000), 'Fashion: Yesterday, Today and Tomorrow', in Nicola White and Ian Griffiths (eds), *The Fashion Business: Theory, Practice, Image*, Oxford: Berg, pp. 7–20.

Sun, Liping (2006), 'China Leapfrogs into a "Digital Music Age": International Recording Industry Giants Converge at Music Fair', *Xinhua Net* [Online], 16 May, available at http://news.xinhuanet.com/newmedia/2006-05/17/content_4555754.htm (accessed 4 February 2010) (in Chinese).

Sun, Shaoyi (2000), 'Under the Shadow of Commercialization: The Changing Landscape of Chinese Cinema', *Celluloid*, April, 1–6.

Tabarrok, Alexander (2002), 'Patent Theory versus Patent Law', *B.E. Journal of Economic Analysis and Policy*, 1 (10) [Online], available at http://www.bepress.com/bejeap/contributions/vol1/iss1/art9/ (accessed 2 February 2010).

Towse, Ruth (2001), *Creativity, Incentive and Reward*, Cheltenham, UK and Northampton, MA, USA: Edward Elgar Publishing.

Treverton, Gregory, Carl Matthies, Karla Cunningham, Jeremiah Goulka, Greg Ridgeway and Anny Wong (2009), *Film Piracy, Organised Crime, and Terrorism*, Santa Monica, CA: RAND Corporation.

UK Trade and Investment (2004), 'Changing China – The Creative Industry Perspective: A Market Analysis of China's Digital and Design Industries', available at http://www.eastmids-china.co.uk/uploads/creativeindustry.pdf (accessed 4 February 2010).

UNESCO (2005), 'International Flows of Selected Cultural Goods and Services 1994–2003', available at www.uis.unesco.org/template/pdf/cscl/IntlFlows_EN.pdf (accessed 4 February 2010).

van Schijndel, Marieke and Joost Smiers (2005), 'Imagine a World without Copyright', *International Herald Tribune* [Online], 8 October, available at http://www.iht.com/articles/2005/10/07/opinion/edsmiers.php (accessed 3 February 2010).

Veblen, Thorstein (1899), *The Theory of the Leisure Class: An Economic Study of Institutions*, reprinted in Michael Lewis (ed.) (2007), *The Real Price of Everything: Rediscovering the Six Classics of Economics*, New York: Sterling Publishing.

Wan, Jihong and Richard Kraus (2002), 'Hollywood and China as Adversaries and Allies', *Public Affairs*, 75 (3), 419–34.

Wang, Jing (2004), 'The Global Reach of a New Discourse: How Far Can "Creative Industries" Travel?', *International Journal of Cultural Studies*, 7 (1), 9–19.

Wang, Shujen (2003), *Framing Piracy*, Lanham, MD: Rowman & Littlefield.

Wang, Yumei (2008), 'Sun Shoushan: Our Country Has Authorized 59 Chinese and Foreign Periodical Cooperative Projects', *Xinhua Net* [Online], 3 April, available at http://news.xinhuanet.com/newmedia/2008-04/03/content_7913167.htm (accessed 3 February 2010) (in Chinese).

Wood, Hannah (2009), 'Fake Brands Shopping Centre Set to Open in China – Pictures', *Mirror* [Online], 5 January, available at http://www.mirror.co.uk/news/top-stories/2009/01/05/fake-brands-shopping-centre-set-to-open-in-china-pictures-115875-21018152/ (accessed 7 February 2010).

World Economic Forum (2008), 'China Must Take Steps to Boost Domestic Consumption', Press Release [Online], 28 September, available at http://www.weforum.org/en/media/Latest%20Press%20Releases/PR_AMNC08_Cons (accessed 4 February 2010).

Wu, Juanjuan (2009), *Chinese Fashion: From Mao to Now*, Oxford and New York: Berg.

Xu, Gary (2007), *Sinascape: Contemporary Chinese Cinema*, Lanham, MD and Plymouth: Rowman & Littlefield.

Xue, Hong and Zheng Chengshi (2002), *Chinese Intellectual Property Law in the 21st Century*, Hong Kong, Singapore and Petaling Jaya, Malaysia: Sweet & Maxwell Asia.

Yan, Jiaqi and Gao Gao (1996), *Turbulent Decade: A History of the Cultural Revolution*, translated and edited by D.W.Y. Kwok, Honolulu: University of Hawai'i Press.

Yao, C. (2007), 'A Prosperous Market of Wireless Music', *Communicate*, 1 (28) [Online], available at http://www.huawei.com/publications/view.do?id=1621&cid=3162&pid=61 (accessed 3 February 2010).

Yung, Danny (2003), 'Creative Industry, Creative Hong Kong Forum', available at http://www.cpu.gov.hk/english/documents/conference/20030924 Opening%20Speech.pdf (accessed 4 February 2010).

Yurchak, Alexei (2002), 'Entrepreneurial Governmentality in Post-Socialist Russia: A Cultural Investigation of Business Practices', in Victoria Bonnell and Thomas Gold (eds), *The New Entrepreneurs of Europe and Asia*, New York: M.E. Sharpe, pp. 278–324.

Zhang, Yingjin (2004), *Chinese National Cinema*, New York: Routledge.

LIST OF INTERVIEWS

Chen, Daming (2004), director and producer, interview with the author, Beijing, 18 March.

Chen, Xiang (2009), Shanghai Shiyi Company Limited, director, interview with the author, Shanghai, 27 April.

Hu, Bo (2004), Beijing Longevity Digital Entertainment Production Company, director, producer, interview with the author, Beijing, 15 July.

Huang, Beaker (2005), Warner Music China, marketing and business development director, interview with the author, Beijing, 10 August.

Huang, Qunfei (2009), New Film Association, general manager, interview with the author, Beijing, 1 April.

Hung, Huang (2005), China Interactive Media Group, chief executive officer, interview with the author, Beijing, 12 August.

Jia, Qi (2009), Academy of Fine Arts, Harbin Normal University, associate director, interview with the author, 13 April, Beijing.

Kuo, Kaiser (2007), Ogilvy, group director for digital media, interview with the author, Beijing, 31 October.

Lan, Simon (2005), Beijing Frontline Productions, producer, interview with the author, Beijing, 24 June.

Li, Wei (2009), Tsinghua University School of Art and Design, associate professor, interview with the author, Beijing, 27 March.

Li, Yang (2004), Tang Splendour Film Company, director, producer, interview with the author, Beijing, 5 May.

Liu, Xinxin (2009), Tsinghua University School of Art and Design, lecturer and independent artist, interview with the author, Beijing, 20 March.

Tang, Terry (2007), Noank Media, director of China business development, interview with the author, Beijing, 2 November.

Wang, Jade (2005), deputy channel director, Easy FM, China Radio International, interview with the author, Beijing, 2 August.

Wang, Shi (2004), Chinese Culture Protection Society, interview with the author, Beijing, 1 August.

Wang, Zhebin and Wang, Xingdong (2004), Beijing Forbidden City Film Company: Grade A screenwriter, film producer; and Grade A screenwriter, president Script Writers Guild of China, delegate Chinese People's Consultative Conference, interview with the author, Beijing, 8 May.

Wulan, Tana (2004), Xian Film Studios, director, interview with the author, Beijing, 15 July.

Zhao, Daniel (2005), Huayi Music, vice general manager, interview with the author, Beijing, 9 August.

Zhu, Baixi (2009), Fougere 2, co-founder and designer, interview with the author, Shanghai, 18 April.

Index